Problems Supplement to Accompany

Vector Mechanics for Engineers: Statics

Problems Supplement to Accompany

Vector Mechanics for Engineers

Statics

Fifth Edition

Ferdinand P. Beer
Lehigh University

E. Russell Johnston, Jr.
University of Connecticut

McGraw-Hill, Inc.
New York St. Louis San Francisco Auckland Bogotá Caracas
Hamburg Lisbon London Madrid Mexico Milan Montreal
New Delhi Oklahoma City Paris San Juan São Paulo Singapore
Sydney Tokyo Toronto

This book is printed on recycled paper.

PROBLEMS SUPPLEMENT TO ACCOMPANY
VECTOR MECHANICS FOR ENGINEERS: Statics

1 2 3 4 5 6 7 8 9 0 MAL MAL 9 0 9 8 7 6 5 4 3 2

ISBN 0-07-005011-2

The editor was John J. Corrigan;
the production supervisor was Phil Galea.
Malloy Lithographing, Inc., was printer and binder.

Contents

CHAPTER 2
STATICS OF PARTICLES

Answers to all problems are given at the end of the booklet.

SECTIONS 2.1 TO 2.6

2.1 and 2.2 Determine graphically the magnitude and direction of the resultant of the two forces shown, using in each problem (*a*) the parallelogram law, (*b*) the triangle rule.

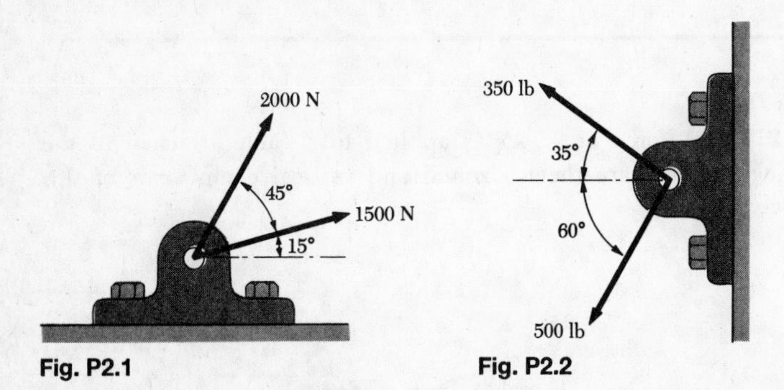

Fig. P2.1

Fig. P2.2

2.3 Two structural members *B* and *C* are riveted to the bracket *A*. Knowing that the tension in member *B* is 2500 lb and that the tension in *C* is 2000 lb, determine graphically the magnitude and direction of the resultant force acting on the bracket.

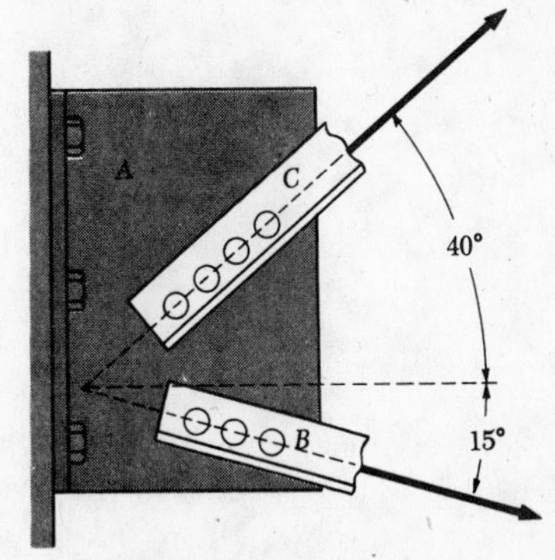

Fig. P2.3 and P2.4

2.4 Two structural members *B* and *C* are riveted to the bracket *A*. Knowing that the tension in member *B* is 6 kN and that the tension in *C* is 10 kN, determine graphically the magnitude and direction of the resultant force acting on the bracket.

2.5 The force **F** of magnitude 100 lb is to be resolved into two components along the lines *a-a* and *b-b*. Determine by trigonometry the angle α, knowing that the component of **F** along the line *a-a* is to be 70 lb.

2.6 The force **F** of magnitude 800 N is to be resolved into two components along the lines *a-a* and *b-b*. Determine by trigonometry the angle α, knowing that the component of **F** along the line *b-b* is to be 120 N.

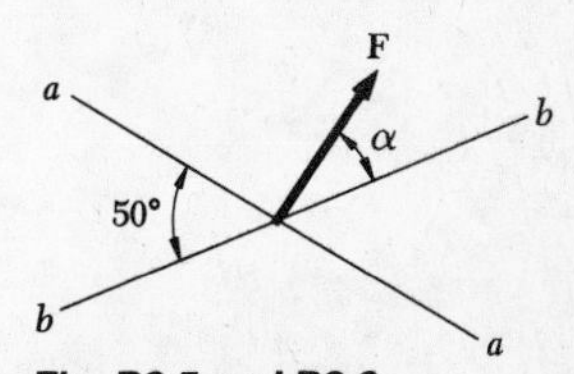

Fig. P2.5 and P2.6

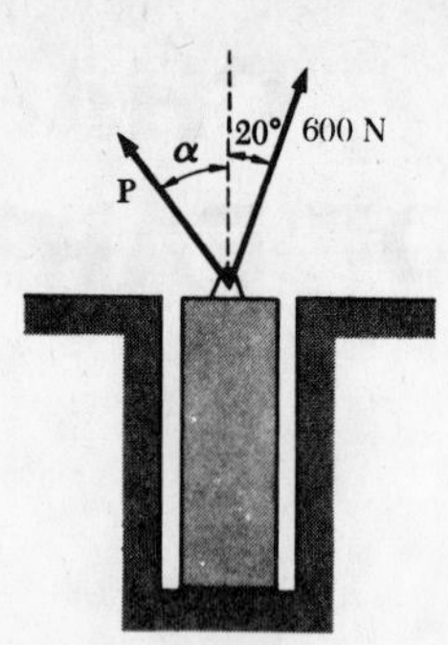

Fig. P2.7 and P2.9

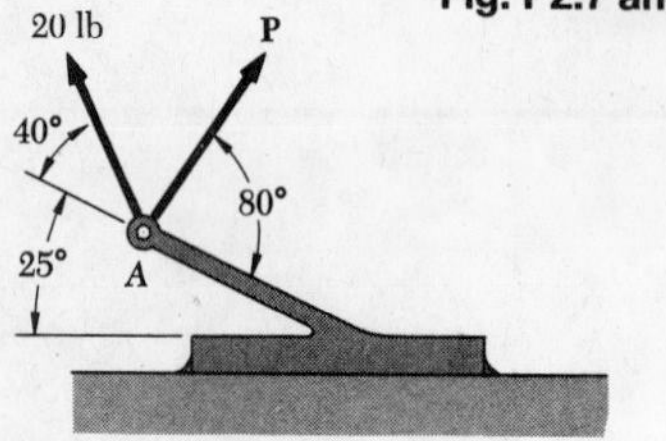

Fig. P2.8 and P2.10

2.7 Knowing that $\alpha = 30°$, determine the magnitude of the force **P** so that the resultant force exerted on the cylinder is vertical. What is the corresponding magnitude of the resultant?

2.8 Determine by trigonometry the magnitude of the force **P** so that the resultant of the two forces applied at A is vertical. What is the corresponding magnitude of the resultant?

2.9 A cylinder is to be lifted by two cables. Knowing that the tension in one cable is 600 N, determine the magnitude and direction of the force **P** so that the resultant is a vertical force of 900 N.

2.10 Knowing that $P = 30$ lb, determine by trigonometry the resultant of the two forces applied at point A.

2.11 Solve Prob. 2.3 by trigonometry.

2.12 If the resultant of the two forces acting on the cylinder of Prob. 2.7 is to be vertical, find (*a*) the value of α for which the magnitude of **P** is minimum, (*b*) the corresponding magnitude of **P**.

SECTIONS 2.7 AND 2.8

2.13 A force of 2.5 kN is applied to a cable attached to the bracket. What are the horizontal and vertical components of this force?

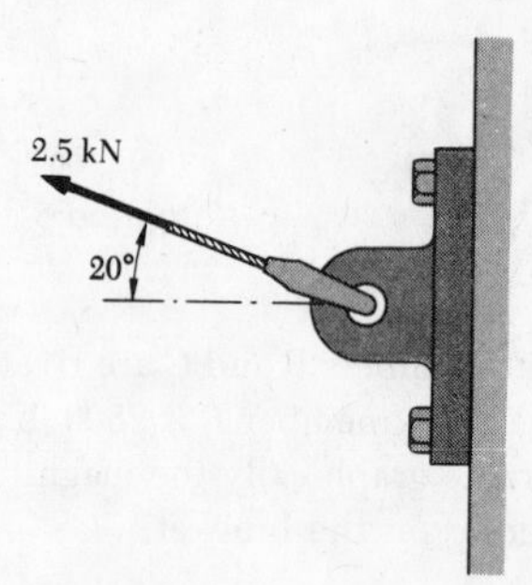

Fig. P2.13

2.14 The force **P** must have a 60-lb component acting up the incline. Determine the magnitude of **P** and of its component perpendicular to the incline.

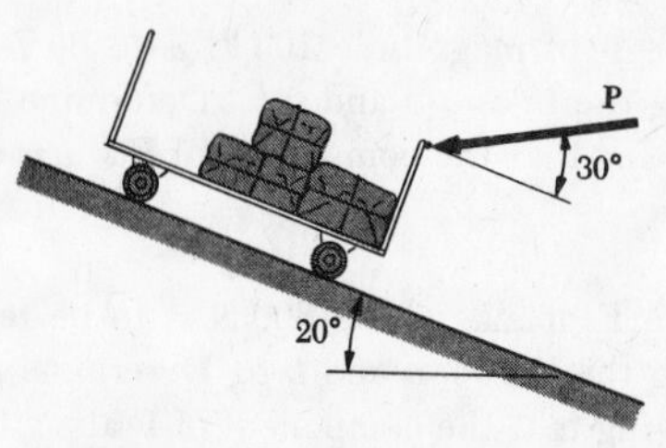

Fig. P2.14

2.15 The tension in the support wire AB is 65 lb. Determine the horizontal and vertical components of the force acting on the pin at A.

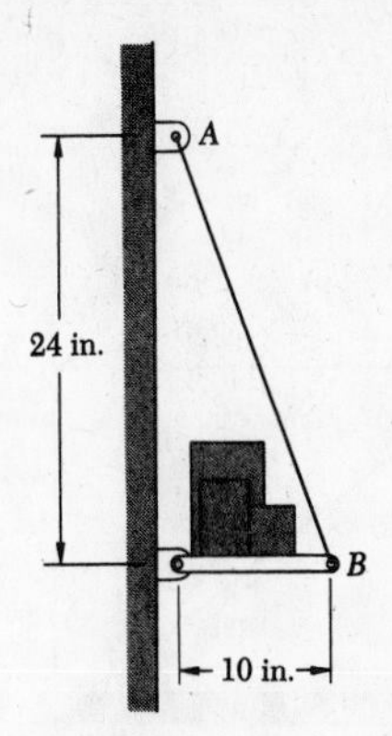

Fig. P2.15

2.16 The hydraulic cylinder GE exerts on member DF a force **P** directed along line GE. Knowing that **P** must have a 600-N component perpendicular to member DF, determine the magnitude of **P** and of its component parallel to DF.

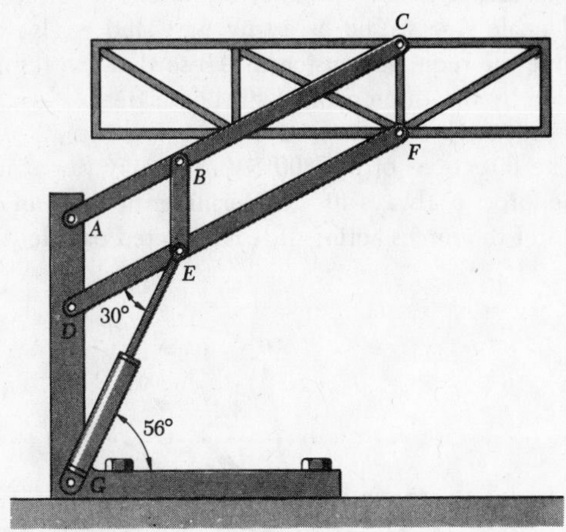

Fig. P2.16

2.17 Determine the x and y components of each of the forces shown.

2.18 Determine the x and y components of each of the forces shown.

2.19 Determine separately the components of the 80-N force and of the 120-N force in directions parallel and perpendicular to the line of action of the 100-N force.

2.20 Determine separately the components of the 40-lb force and of the 60-lb force in directions parallel and perpendicular to the line of action of the 50-lb force.

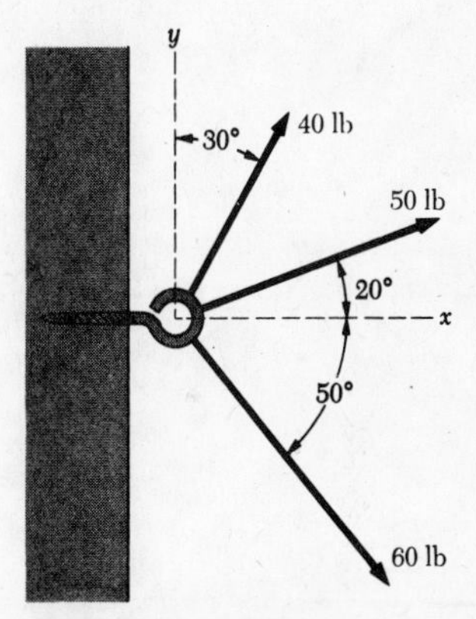

Fig. P2.17 and P2.20

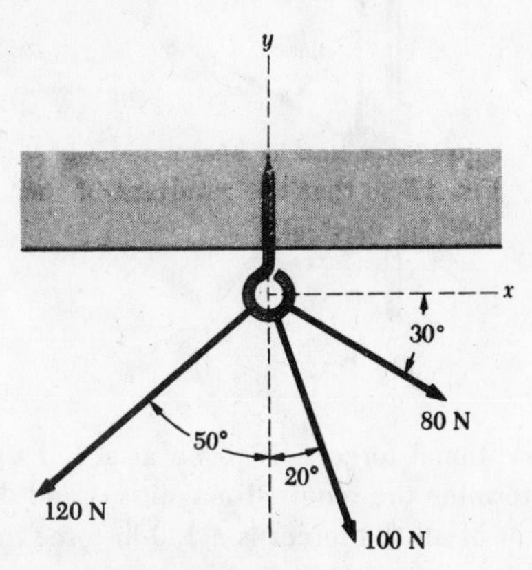

Fig. P2.18 and P2.19

2.21 Determine the resultant of the three forces of Prob. 2.18.

2.22 Determine the resultant of the three forces of Prob. 2.17.

2.23 Using x and y components, solve Prob. 2.3.

2.24 Using x and y components, solve Prob. 2.1.

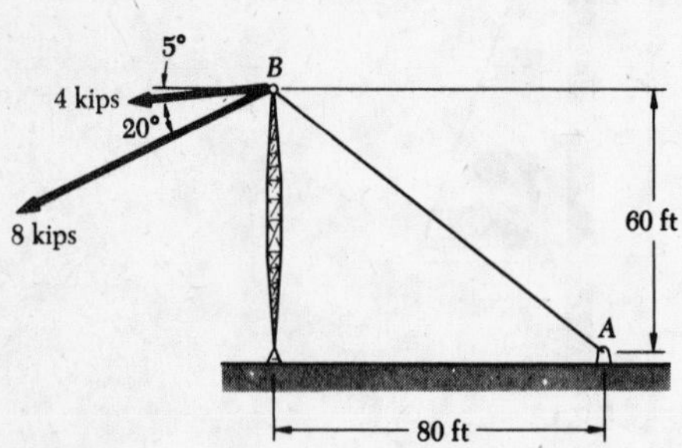

Fig. P2.25

2.25 Two cables which have known tensions are attached at point B. A third cable AB is used as a guy wire and is also attached at B. Determine the required tension in AB so that the resultant of the forces exerted by the three cables will be vertical.

2.26 The directions of the 300-N forces may vary, but the angle between the forces is always 40°. Determine the value of α for which the resultant of the forces acting at A is directed parallel to the plane b-b.

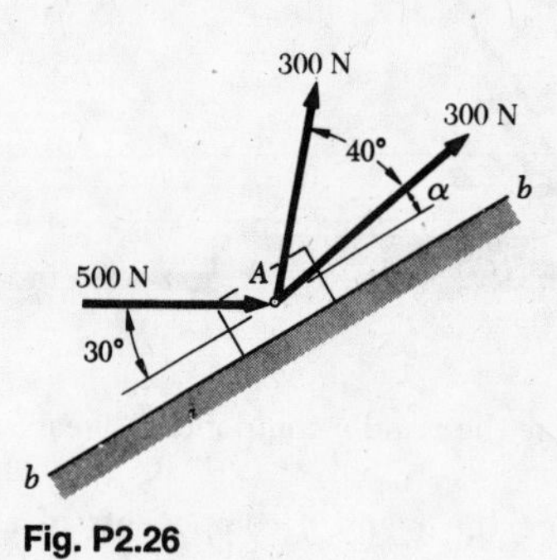

Fig. P2.26

2.27 A collar which may slide on a vertical rod is subjected to the three forces shown. The direction of the force **F** may be varied. If possible, determine the direction of the force **F** so that the resultant of the three forces is horizontal, knowing that the magnitude of **F** is (*a*) 2.4 kN, (*b*) 1.4 kN.

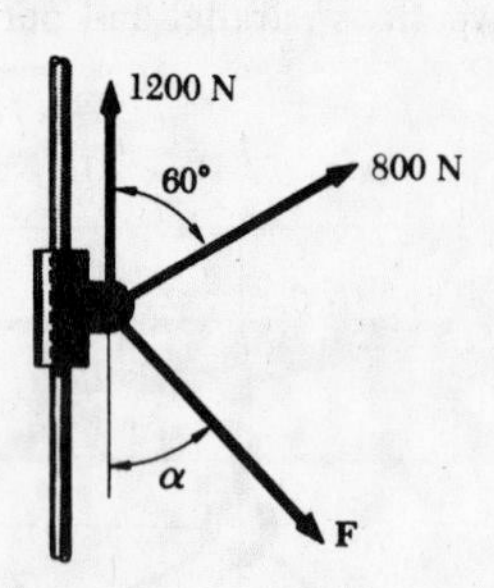

Fig. P2.27

2.28 An additional force **F** is to be attached to the eyebolt of Prob. 2.17. Determine the required magnitude and direction of **F** so that the resultant of all the forces is a 120-lb force directed horizontally to the right.

SECTIONS 2.9 TO 2.11

2.29 through 2.32 Two cables are tied together at C and loaded as shown. Determine the tension in AC and BC.

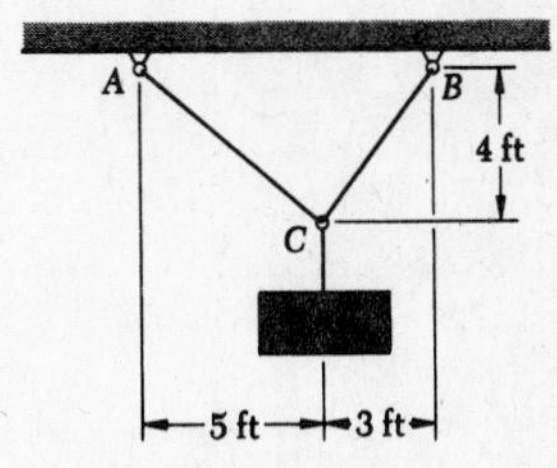

Fig. P2.29

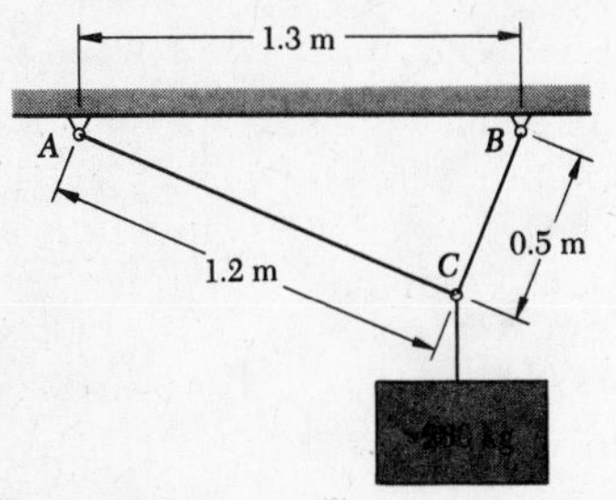

Fig. P2.30

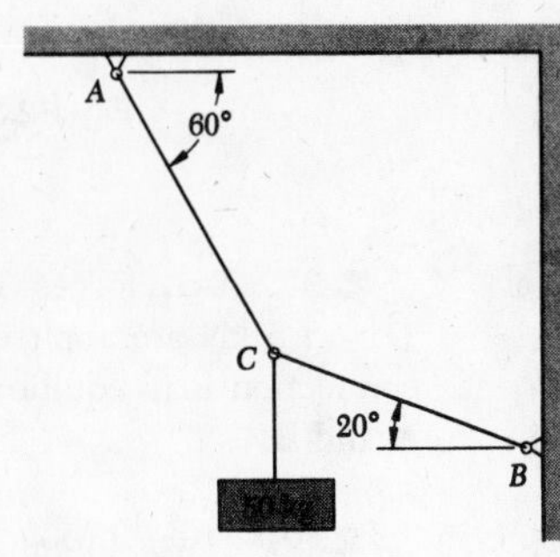

Fig. P2.31

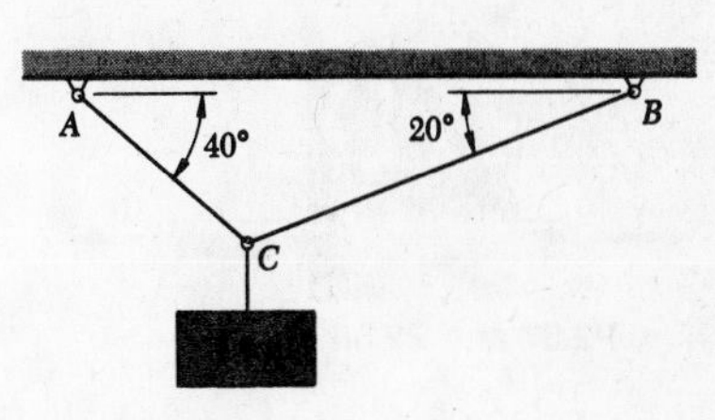

Fig. P2.32

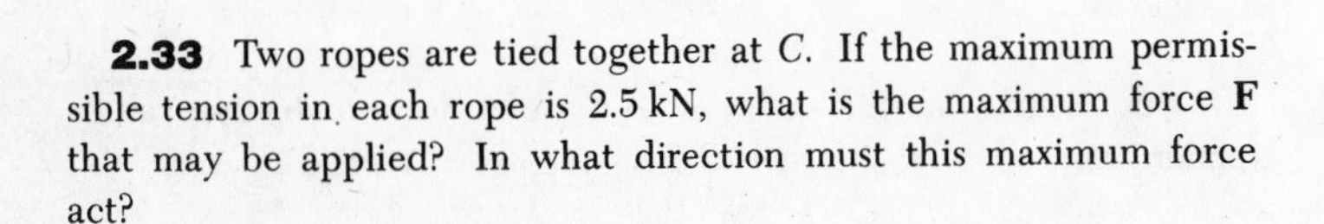

2.33 Two ropes are tied together at C. If the maximum permissible tension in each rope is 2.5 kN, what is the maximum force $\mathbf{F}$ that may be applied? In what direction must this maximum force act?

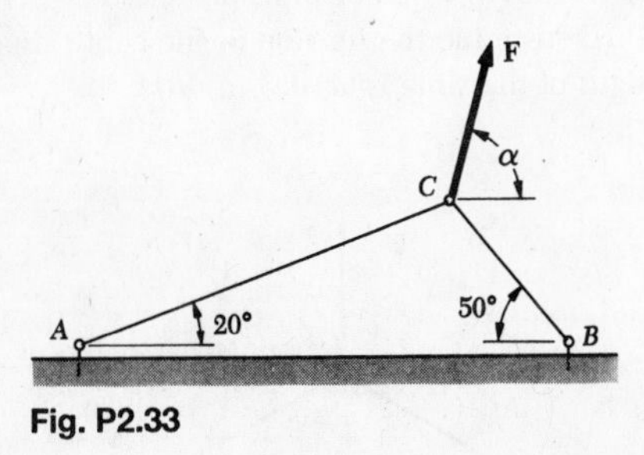

Fig. P2.33

2.34 A 600-lb block is supported by the two cables AC and BC. (a) For what value of α is the tension in cable AC minimum? (b) What are the corresponding values of the tension in cables AC and BC?

2.35 A 600-lb block is supported by the two cables AC and BC. Determine (a) the value of α for which the larger of the cable tensions is as small as possible, (b) the corresponding values of the tension in cables AC and BC.

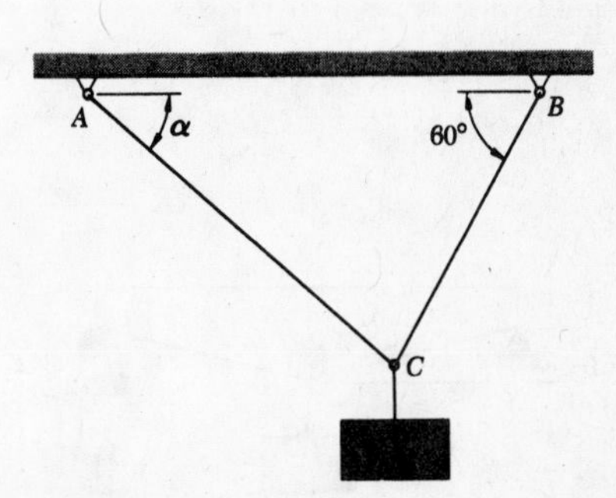

Fig. P2.34 and P2.35

2.36 The force **P** is applied to a small wheel which rolls on the cable ACB. Knowing that the tension in both parts of the cable is 750 N, determine the magnitude and direction of **P**.

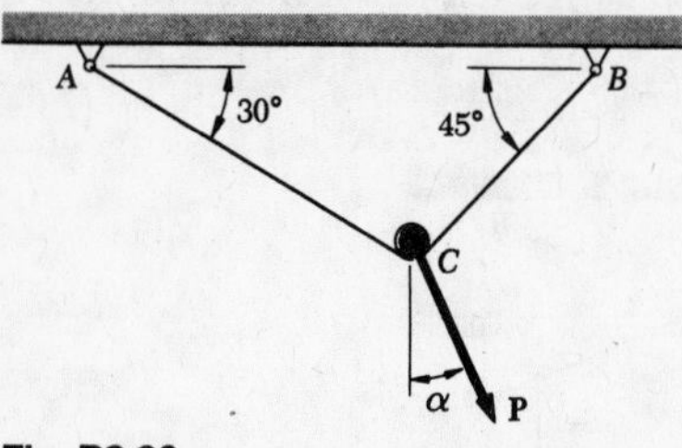

Fig. P2.36

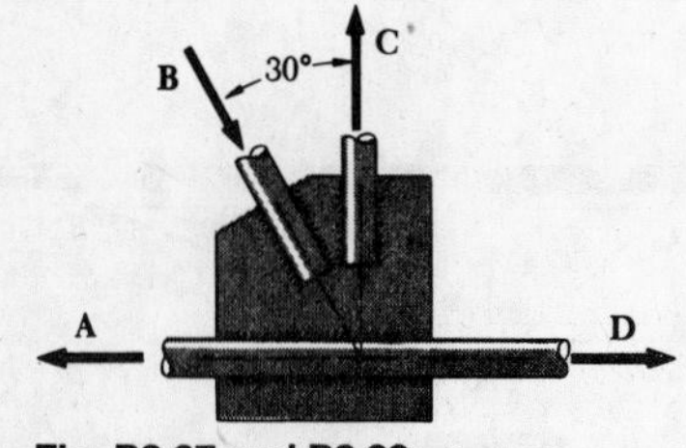

Fig. P2.37 and P2.38

2.37 Two forces **C** and **D** of magnitude $C = 4$ kN and $D = 7.5$ kN are applied to the connection shown. Knowing that the connection is in equilibrium, determine the magnitudes of the forces **A** and **B**.

2.38 Two forces **A** and **B** of magnitude $A = 5000$ N and $B = 2500$ N are applied to the connection shown. Knowing that the connection is in equilibrium, determine the magnitudes of the forces **C** and **D**.

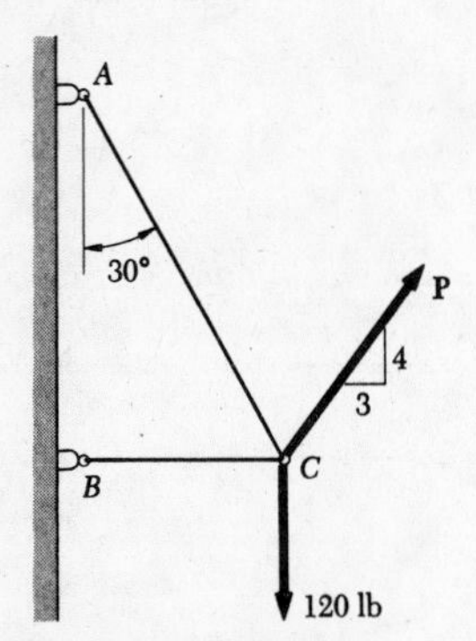

Fig. P2.39 and P2.40

2.39 Knowing that $P = 100$ lb, determine the tension in cables AC and BC.

2.40 Determine the range of values of **P** for which both cables remain taut.

2.41 A 3.6-m length of steel pipe of mass 300 kg is lifted by a crane cable CD. Determine the tension in the cable sling ACB, knowing that the length of the sling is (*a*) 4.5 m, (*b*) 6 m.

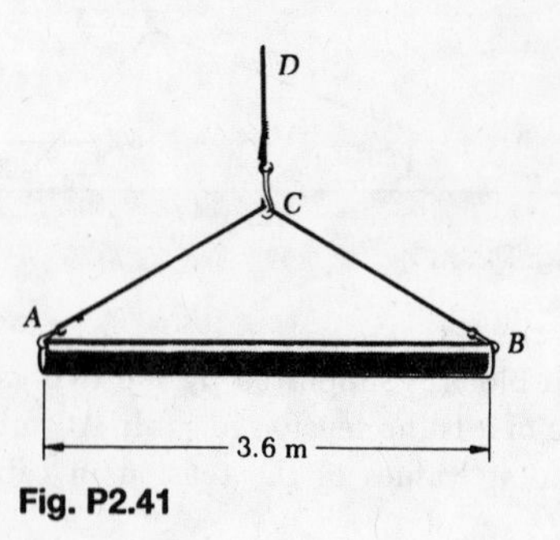

Fig. P2.41

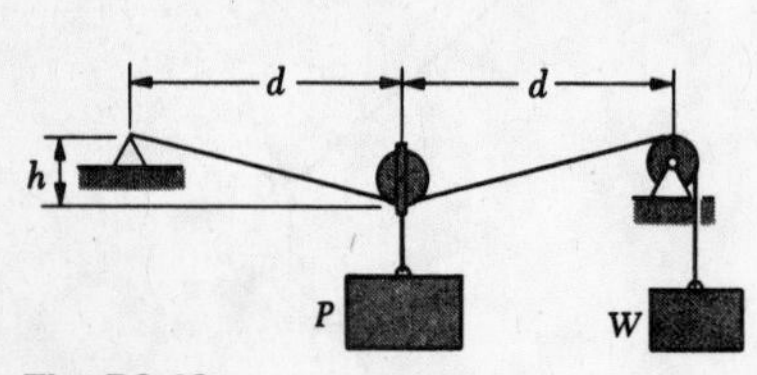

Fig. P2.42

2.42 If in the diagram shown $W = 80$ lb, $P = 10$ lb, and $d = 20$ in., determine the value of h consistent with equilibrium.

2.43 The collar A may slide freely on the horizontal smooth rod. Determine the magnitude of the force $\mathbf{P}$ required to maintain equilibrium when (*a*) $c = 9$ in., (*b*) $c = 16$ in.

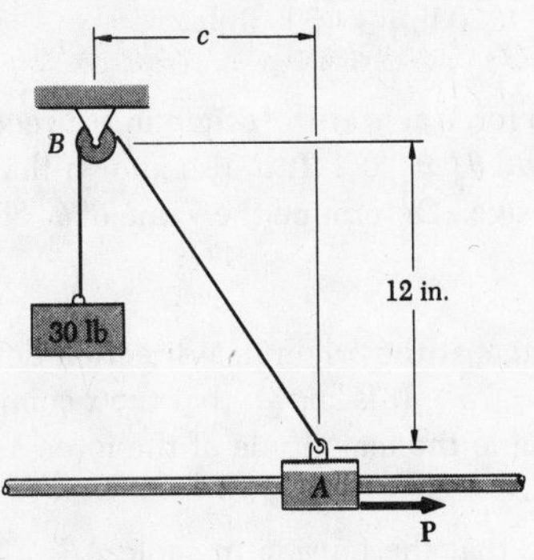

Fig. P2.43

***2.44** A 150-kg block is attached to a small pulley which may roll on the cable ACB. The pulley and load are held in the position shown by a second cable DE which is parallel to the portion CB of the main cable. Determine (*a*) the tension in cable ACB, (*b*) the tension in cable DE. Neglect the radius of the pulleys and the weight of the cables.

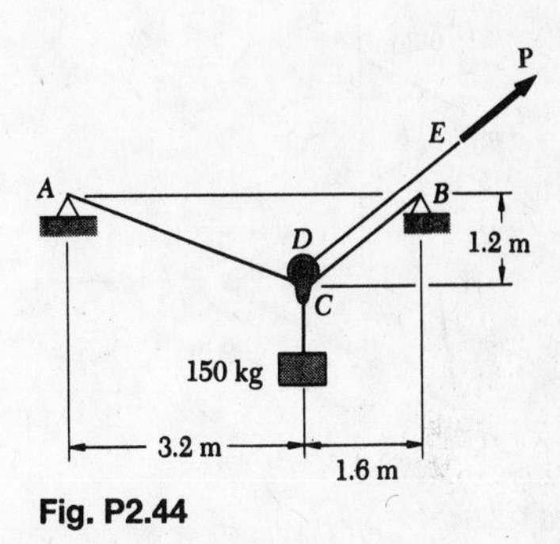

Fig. P2.44

SECTIONS 2.12 TO 2.14

2.45 Determine (*a*) the x, y, and z components of the 250-N force, (*b*) the angles θ_x, θ_y, and θ_z that the force forms with the coordinate axes.

2.46 Determine (*a*) the x, y, and z components of the 300-N force, (*b*) the angles θ_x, θ_y, and θ_z that the force forms with the coordinate axes.

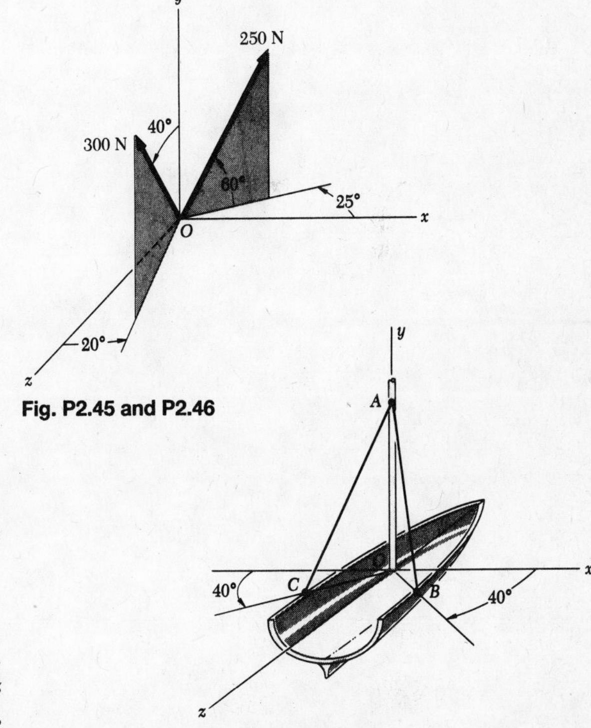

Fig. P2.45 and P2.46

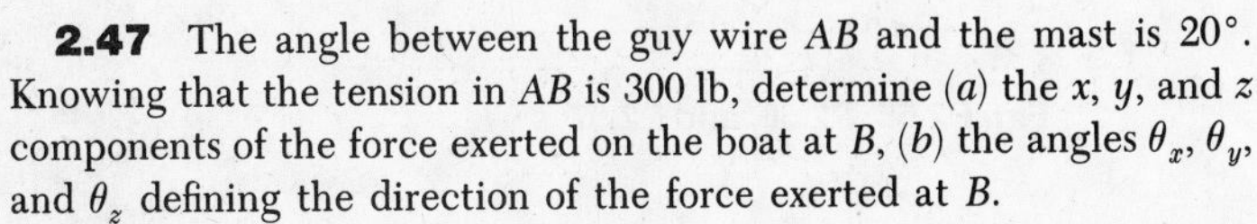

2.47 The angle between the guy wire AB and the mast is 20°. Knowing that the tension in AB is 300 lb, determine (*a*) the x, y, and z components of the force exerted on the boat at B, (*b*) the angles θ_x, θ_y, and θ_z defining the direction of the force exerted at B.

2.48 The angle between the guy wire AC and the mast is 20°. Knowing that the tension in AC is 300 lb, determine (*a*) the x, y, and z components of the force exerted on the boat at C, (*b*) the angles θ_x, θ_y, and θ_z defining the direction of the force exerted at C.

Fig. P2.47 and P2.48

2.49 Determine the magnitude and direction of the force $\mathbf{F} = (700\text{ N})\mathbf{i} - (820\text{ N})\mathbf{j} + (960\text{ N})\mathbf{k}$.

2.50 Determine the magnitude and direction of the force $\mathbf{F} = -(240\text{ lb})\mathbf{i} - (320\text{ lb})\mathbf{j} + (600\text{ lb})\mathbf{k}$.

2.51 A 250-lb force acts at the origin in a direction defined by the angles $\theta_x = 65°$ and $\theta_y = 40°$. It is also known that the z component of the force is positive. Determine the value of θ_z and the components of the force.

2.52 A force acts at the origin in a direction defined by the angles $\theta_y = 120°$ and $\theta_z = 75°$. It is known that the x component of the force is +40 N. Determine the magnitude of the force and the value of θ_x.

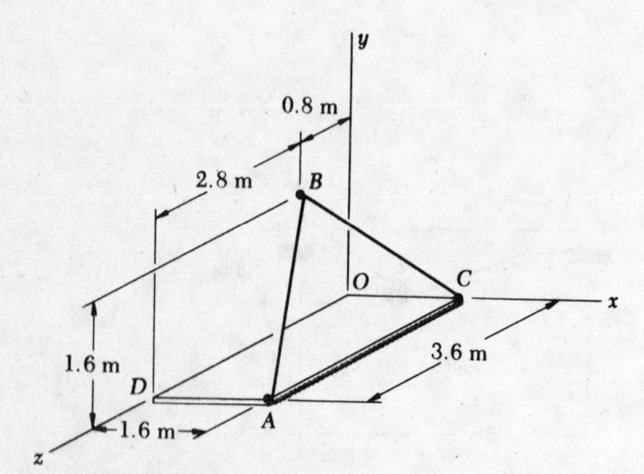

Fig. P2.53 and P2.54

2.53 Knowing that the tension in cable AB is 900 N, determine the components of the force exerted on the plate at A.

2.54 Knowing that the tension in cable BC is 450 N, determine the components of the force exerted on the plate at C.

2.55 Determine the angles θ_x, θ_y, and θ_z which define the direction of the force exerted on point A.

2.56 Determine the angles θ_x, θ_y, and θ_z which define the direction of the force exerted on point C.

2.57 The tension in each of the cables AB and BC is 1350 lb. Determine the components of the resultant of the forces exerted on point B.

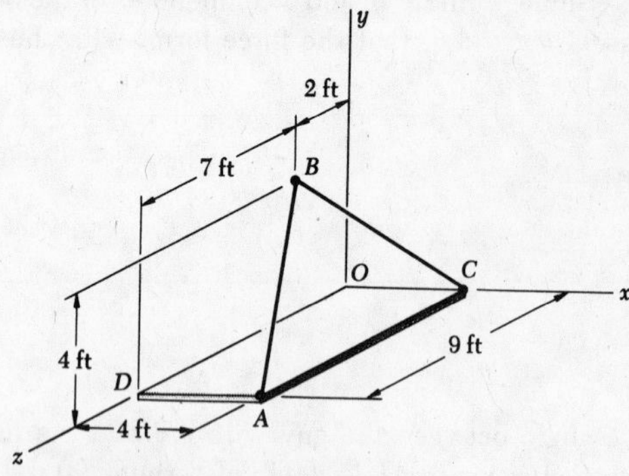

Fig. P2.55, P2.56, and P2.57

2.58 Determine the two possible values of θ_y for a force $\mathbf{F}$, (*a*) if the force forms equal angles with the positive x, y, and z axes, (*b*) if the force forms equal angles with the positive y and z axes and an angle of 45° with the positive x axis.

2.59 Knowing that the tension in AB is 39 kN, determine the required values of the tension in AC and AD so that the resultant of the three forces applied at A is vertical.

2.60 Knowing that the tension in AC is 28 kN, determine the required values of the tension in AB and AD so that the resultant of the three forces applied at A is vertical.

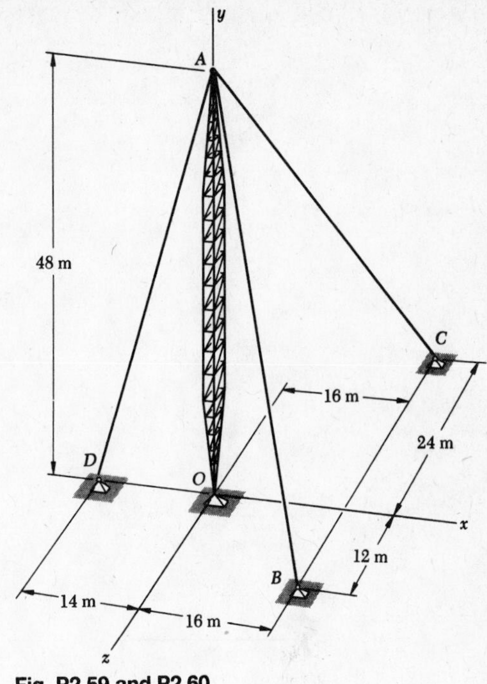

Fig. P2.59 and P2.60

SECTION 2.15

2.61 Three cables are joined at D where two forces $\mathbf{P} = (4 \text{ kN})\mathbf{i}$ and $\mathbf{Q} = 0$ are applied. Determine the tension in each cable.

2.62 Three cables are joined at D where two forces $\mathbf{P} = (3.5 \text{ kN})\mathbf{i}$ and $\mathbf{Q} = (1.5 \text{ kN})\mathbf{k}$ are applied. Determine the tension in each cable.

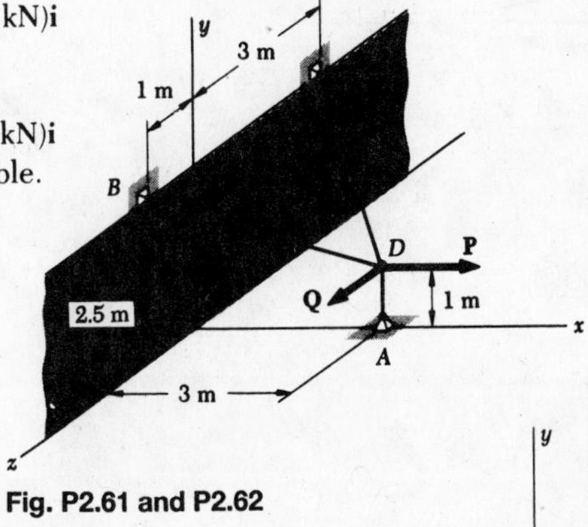

Fig. P2.61 and P2.62

2.63 A load W is supported by three cables as shown. Determine the value of W, knowing that the tension in cable BD is 975 lb.

2.64 A load W is supported by three cables as shown. Determine the value of W, knowing that the tension in cable CD is 300 lb.

2.65 A load W of magnitude 555 lb is supported by three cables as shown. Determine the tension in each cable.

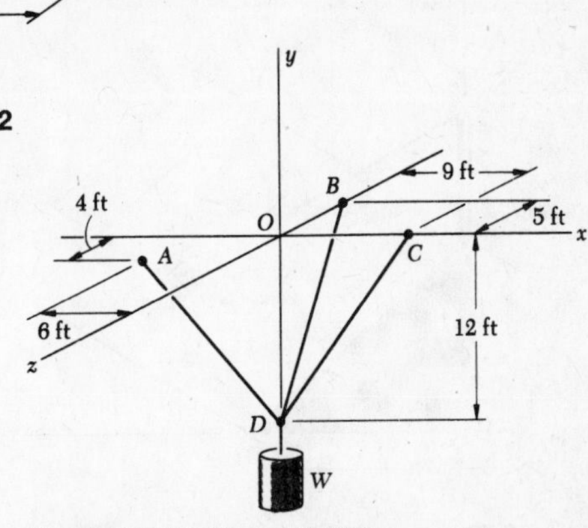

Fig. P2.63, P2.64, and P2.65

2.66 In addition to the 15.60-kN force shown, a force **P** is applied at D in a direction parallel to the y axis. Determine the required magnitude and sense of **P** if the tension in cable CD is to be zero.

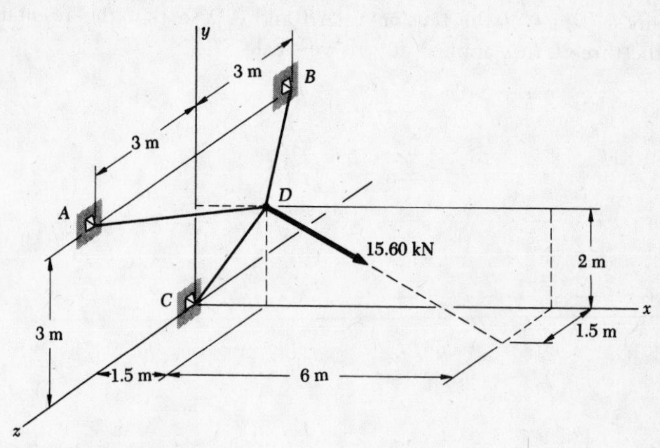

Fig. P2.66 and P2.67

2.67 Three cables are connected at D, where a 15.60-kN force is applied as shown. Determine the tension in each cable.

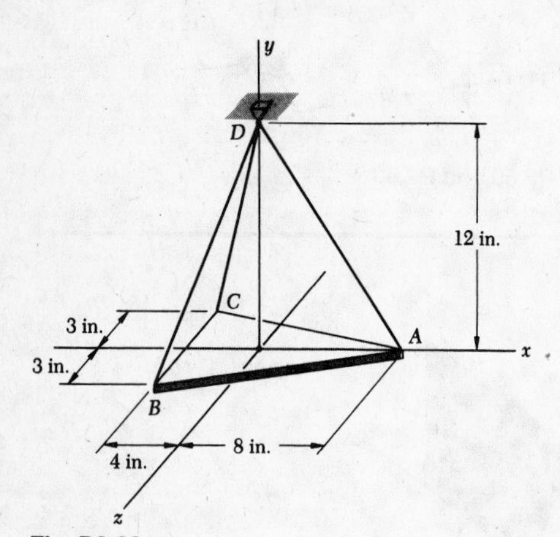

Fig. P2.68

2.68 A triangular plate of weight 18 lb is supported by three wires as shown. Determine the tension in each wire.

2.69 A 15-kg cylinder is supported by three wires as shown. Determine the tension in each wire when $h = 60$ mm.

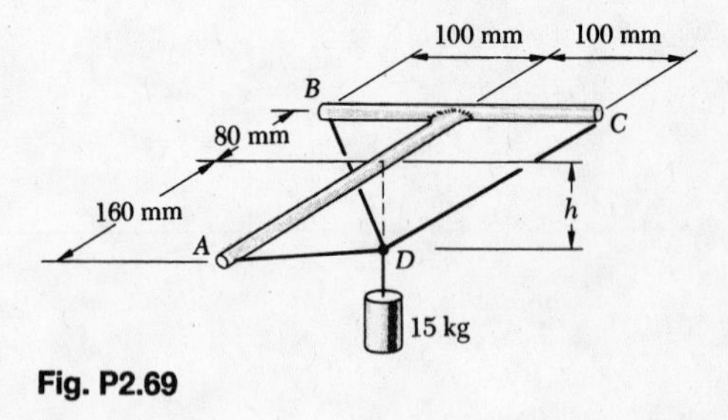

Fig. P2.69

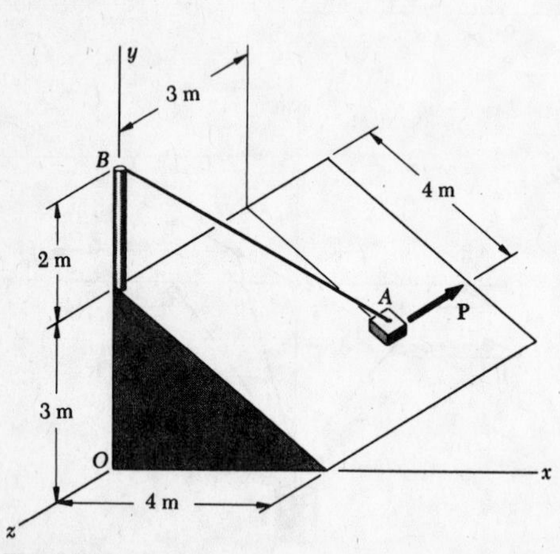

Fig. P2.70

2.70 The 30-kg crate is held on the incline by the wire AB and by the horizontal force **P** which is directed parallel to the z axis. Since the crate is mounted on casters, the force exerted by the incline on the crate is perpendicular to the incline. Determine the magnitude of **P** and the tension in wire AB.

2.71 Collar A weighs 5.6 lb and may slide freely on a smooth vertical rod; it is connected to collar B by wire AB. Knowing that the length of wire AB is 18 in., determine the tension in the wire when (*a*) $c = 2$ in., (*b*) $c = 8$ in.

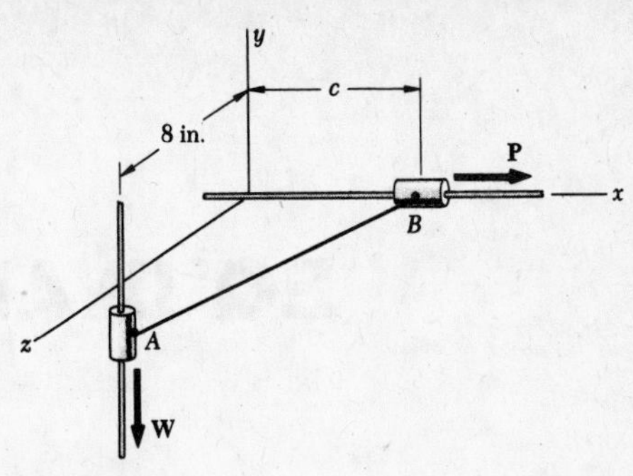

Fig. P2.71

2.72 Solve Prob. 2.71 when (*a*) $c = 14$ in., (*b*) $c = 16$ in.

2.73 The uniform circular ring shown has a mass of 20 kg and a diameter of 300 mm. It is supported by three wires each of length 250 mm. If $\alpha = 120°$, $\beta = 150°$, and $\gamma = 90°$, determine the tension in each wire.

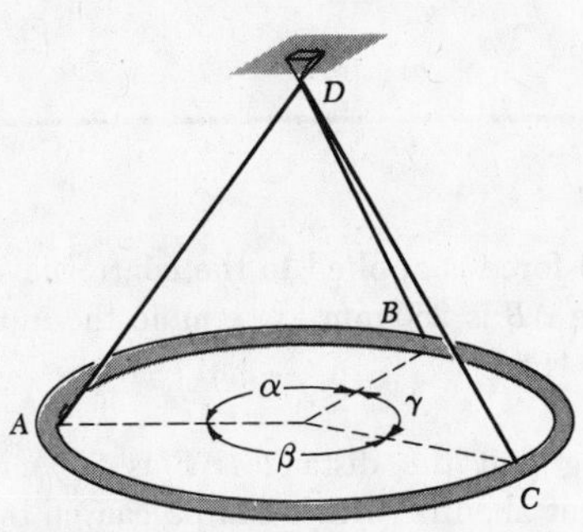

Fig. P2.73

2.74 Three cables are connected at D, where an upward force of 30 kN is applied. Determine the tension in each cable.

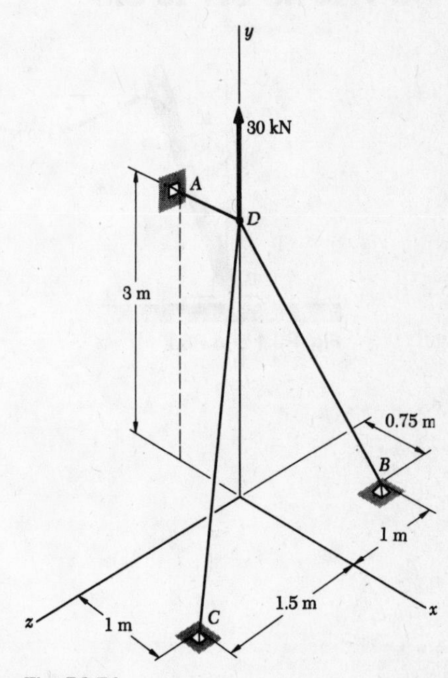

Fig. P2.74

2.75 Two wires are attached to the top of pole CD. It is known that the force exerted by the pole is vertical and that the 500-lb force applied to point C is horizontal. If the 500-lb force is parallel to the z axis ($\alpha = 90°$), determine the tension in each cable.

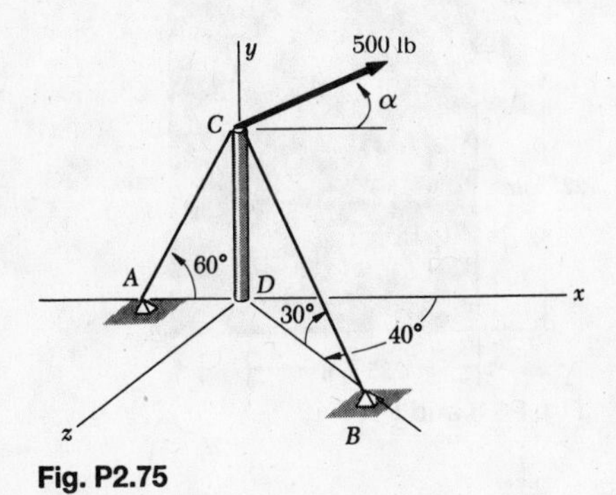

Fig. P2.75

***2.76** In Prob. 2.75, determine (*a*) the value of the angle α for which the tension in cable AC is maximum, (*b*) the corresponding tension in each cable.

CHAPTER 3 RIGID BODIES: EQUIVALENT SYSTEMS OF FORCES

SECTIONS 3.1 to 3.8

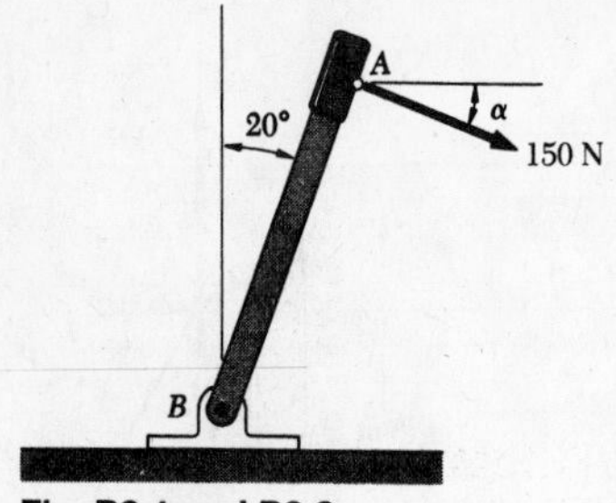

Fig. P3.1 and P3.2

3.1 A 150-N force is applied to the control lever at A. Knowing that the distance AB is 250 mm, determine the moment of the force about B when α is 50°.

3.2 Knowing that the distance AB is 250 mm, determine the maximum moment about B which can be caused by the 150-N force. In what direction should the force act?

3.3 A 450-N force is applied at A as shown. Determine (*a*) the moment of the 450-N force about D, (*b*) the smallest force applied at B which creates the same moment about D.

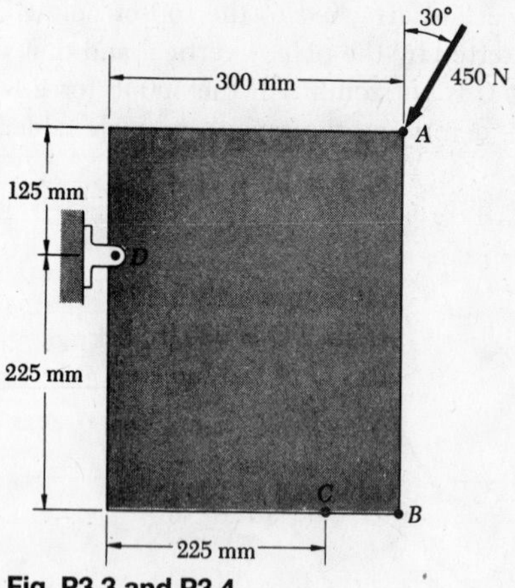

Fig. P3.3 and P3.4

3.4 A 450-N force is applied at A. Determine (*a*) the moment of the 450-N force about D, (*b*) the magnitude and sense of the horizontal force applied at C which creates the same moment about D, (*c*) the smallest force applied at C which creates the same moment about D.

3.5 and 3.6 Compute the moment of the 100-lb force about A, (a) by using the definition of the moment of a force, (b) by resolving the force into horizontal and vertical components, (c) by resolving the force into components along AB and in the direction perpendicular to AB.

3.7 and 3.8 Determine the moment of the 100-lb force about C.

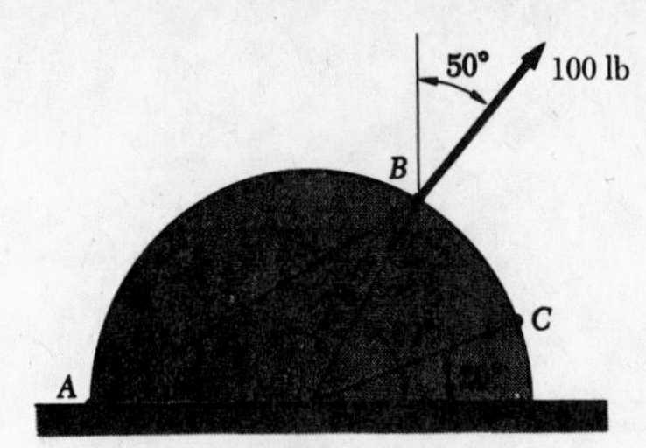

Fig. P3.5 and P3.8

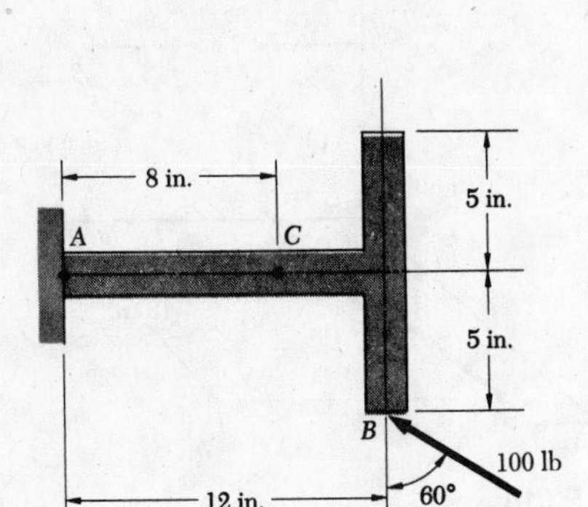

Fig. P3.6 and P3.7

3.9 In Prob. 3.7, determine the perpendicular distance from C to the line of action of the 100-lb force.

3.10 Determine the moment about the origin of coordinates O of the force $\mathbf{F} = 4\mathbf{i} - 3\mathbf{j} + 2\mathbf{k}$ which acts at a point A. Assume that the position of A is (a) $\mathbf{r} = \mathbf{i} + 5\mathbf{j} + 6\mathbf{k}$, ($b$) $\mathbf{r} = 6\mathbf{i} + \mathbf{j} + 3\mathbf{k}$, ($c$) $\mathbf{r} = 5\mathbf{i} - 4\mathbf{j} + 3\mathbf{k}$.

3.11 Determine the moment about the origin of coordinates O of the force $\mathbf{F} = -\mathbf{i} + 3\mathbf{j} + 5\mathbf{k}$ which acts at a point A. Assume that the position of A is (a) $\mathbf{r} = 2\mathbf{i} - 4\mathbf{j} + \mathbf{k}$, ($b$) $\mathbf{r} = 4\mathbf{i} + 6\mathbf{j} + 10\mathbf{k}$, ($c$) $\mathbf{r} = -3\mathbf{i} + 9\mathbf{j} + 15\mathbf{k}$.

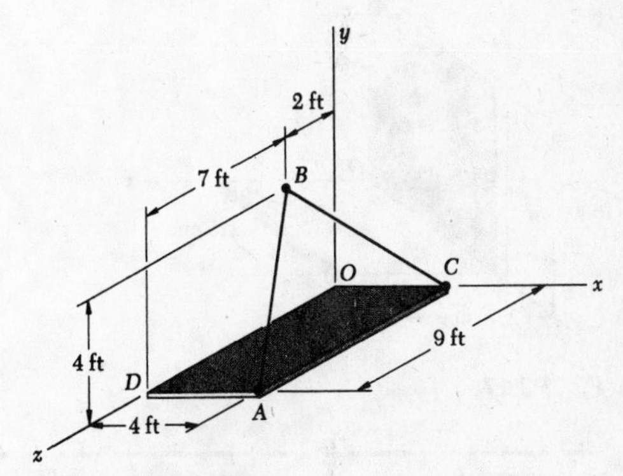

Fig. P3.12 and P3.13

3.12 Knowing that the tension in cable AB is 1800 lb, determine the moment of the force exerted on the plate at A about (a) the origin of coordinates O, (b) corner D.

3.13 Knowing that the tension in cable BC is 900 lb, determine the moment of the force exerted on the plate at C about (a) the origin of coordinates O, (b) corner D.

3.14 A force $\mathbf{P}$ of magnitude 360 N is applied at point B as shown. Determine the moment of $\mathbf{P}$ about (a) the origin of coordinates O, (b) point D.

3.15 A force $\mathbf{Q}$ of magnitude 450 N is applied at point C as shown. Determine the moment of $\mathbf{Q}$ about (a) the origin of coordinates O, (b) point D.

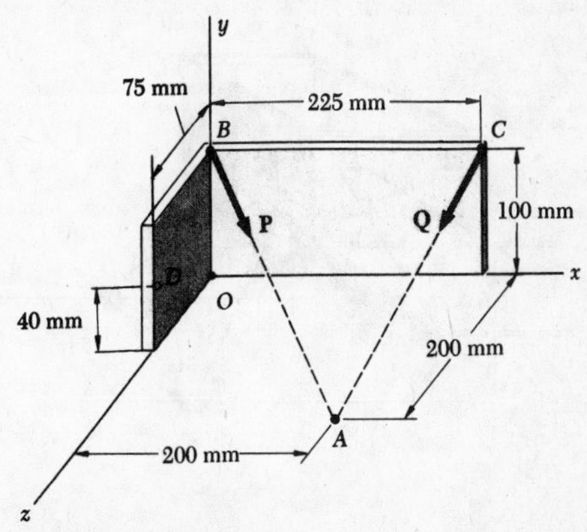

Fig. P3.14 and P3.15

3.16 The line of action of the force **P** of magnitude 420 lb passes through the two points A and B as shown. Compute the moment of **P** about O using the position vector (*a*) of point A, (*b*) of point B.

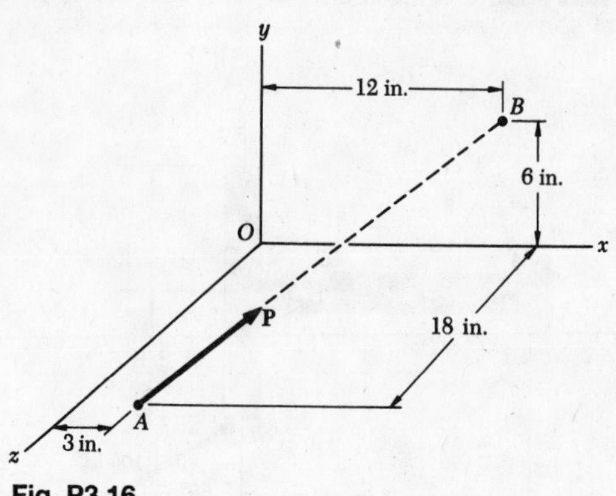

Fig. P3.16

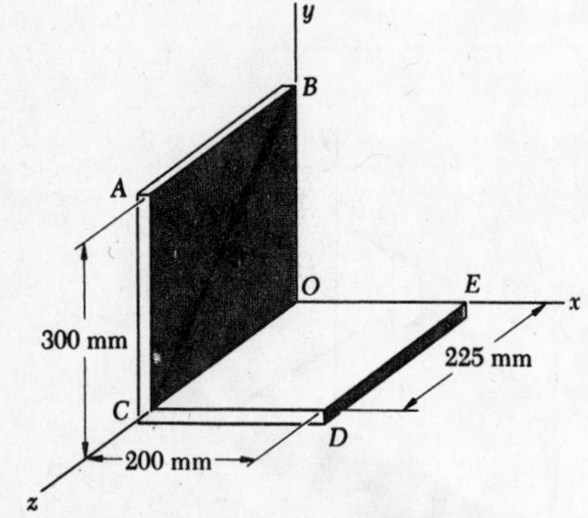

Fig. P3.17

3.17 A force **P** of magnitude 200 N acts along the diagonal BC of the bent plate shown. Determine the moment of **P** about point E.

3.18 In Prob. 3.17, determine the perpendicular distance from the line of action of **P** to point E.

3.19 In Prob. 3.16, determine the perpendicular distance from the line of action of **P** to the origin O.

SECTIONS 3.9 to 3.11

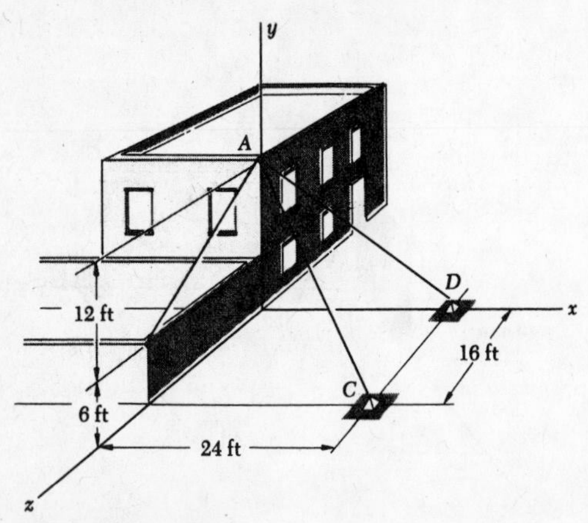

Fig. P3.22 and P3.23

3.20 Given the vectors $\mathbf{P} = 2\mathbf{i} + \mathbf{j} + 2\mathbf{k}$, $\mathbf{Q} = 3\mathbf{i} + 4\mathbf{j} - 5\mathbf{k}$, and $\mathbf{S} = -4\mathbf{i} + \mathbf{j} - 2\mathbf{k}$, compute the scalar products $\mathbf{P} \cdot \mathbf{Q}$, $\mathbf{P} \cdot \mathbf{S}$, and $\mathbf{Q} \cdot \mathbf{S}$.

3.21 Given the vectors $\mathbf{P} = 3\mathbf{i} - 2\mathbf{j} + 6\mathbf{k}$, $\mathbf{Q} = 2\mathbf{i} + 6\mathbf{j} + \mathbf{k}$, and $\mathbf{S} = 5\mathbf{i} + 3\mathbf{j} - 4\mathbf{k}$, compute the scalar products $\mathbf{P} \cdot \mathbf{Q}$, $\mathbf{P} \cdot \mathbf{S}$, and $\mathbf{Q} \cdot \mathbf{S}$.

3.22 Determine the angle formed by (*a*) cables AB and AC, (*b*) cables AB and AD.

3.23 The tension in cable AC is 850 lb. Determine the projection on AB of the force exerted by cable AC at point A.

3.24 The rod ABC consists of a straight segment AB and a circular portion BC. Knowing that $\alpha = 30°$ and that the tension in wire AD is 300 N, determine (*a*) the angle formed by wires AC and AD, (*b*) the projection on AC of the force exerted by wire AD at point A.

3.25 The rod ABC consists of a straight segment AB and a circular portion BC. Knowing that $\alpha = 45°$ and that the tension in wire AC is 450 N, determine (*a*) the angle formed by wires AC and AD, (*b*) the projection on AD of the force exerted by wire AC at point A.

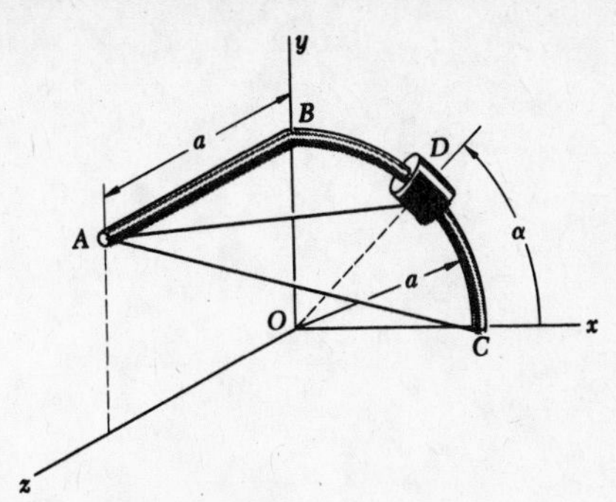

Fig. P3.24 and P3.25

3.26 Knowing that the tension in cable BC is 1400 N, determine (*a*) the angle between cable BC and the boom AB, (*b*) the projection on AB of the force exerted by cable BC at point B.

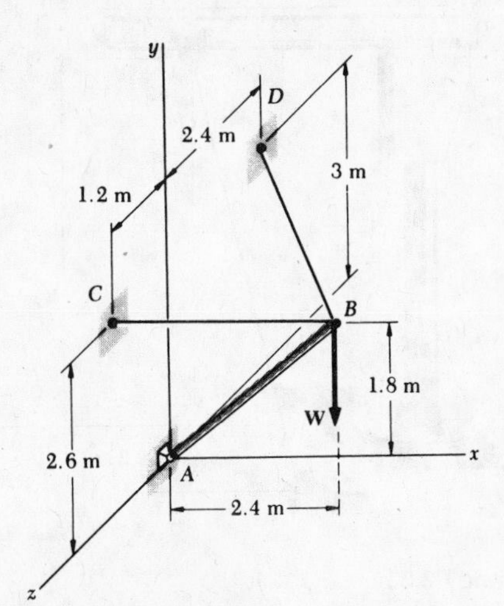

Fig. P3.26 and P3.27

3.27 Knowing that the tension in cable BD is 900 N, determine (*a*) the angle between cable BD and the boom AB, (*b*) the projection on AB of the force exerted by cable BD at point B.

3.28 Knowing that the tension in cable AB is 550 lb, determine (*a*) the angle between cable AB and a line joining points B and C, (*b*) the projection on that line of the force exerted by cable AB at point B.

3.29 Knowing that the tension in cable AB is 550 lb, determine (*a*) the angle formed by AB and AC, (*b*) the projection on AC of the force exerted by cable AB at point A.

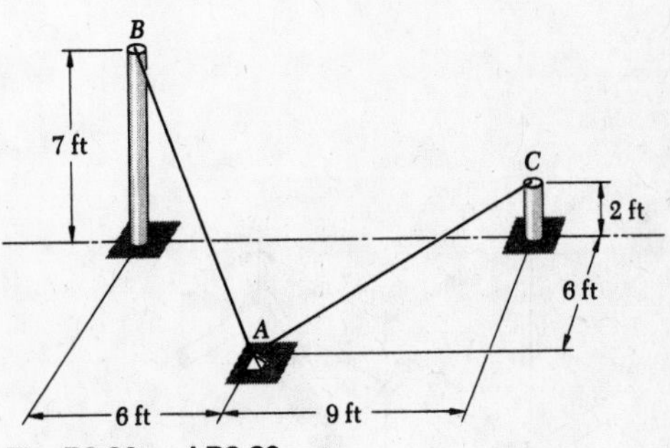

Fig. P3.28 and P3.29

3.30 Given the vectors $\mathbf{P} = 3\mathbf{i} + 2\mathbf{j} + \mathbf{k}$, $\mathbf{Q} = 2\mathbf{i} + \mathbf{j}$, and $\mathbf{S} = \mathbf{i}$, compute $\mathbf{P} \cdot (\mathbf{Q} \times \mathbf{S})$, $(\mathbf{P} \times \mathbf{Q}) \cdot \mathbf{S}$, and $(\mathbf{S} \times \mathbf{Q}) \cdot \mathbf{P}$.

3.31 Given the vectors $\mathbf{P} = 2\mathbf{i} + 3\mathbf{j} + 4\mathbf{k}$, $\mathbf{Q} = -\mathbf{i} + 2\mathbf{j} - 2\mathbf{k}$, and $\mathbf{S} = -3\mathbf{i} - \mathbf{j} + S_z\mathbf{k}$, determine the value of S_z for which the three vectors are coplanar.

3.32 The jib crane is oriented so that the boom DA is parallel to the x axis. At the instant shown the tension in cable AB is 13 kN. Determine the moment about each of the coordinate axes of the force exerted on A by the cable AB.

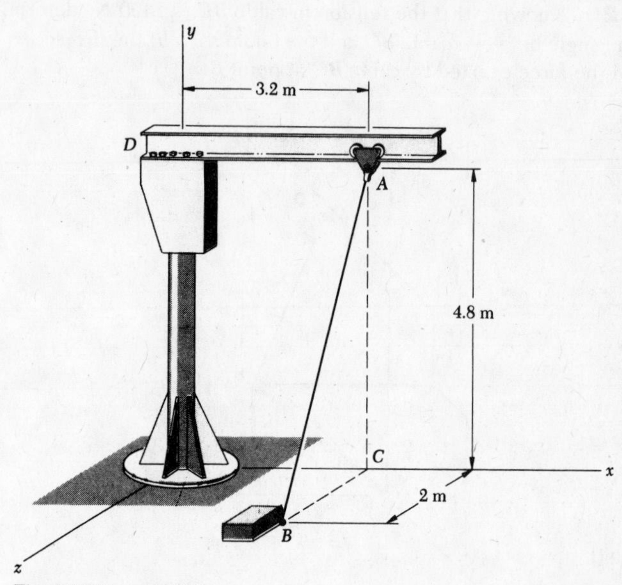

Fig. P3.32 and P3.33

3.33 The jib crane is oriented so that the boom DA is parallel to the x axis. Determine the maximum permissible tension in the cable AB if the absolute values of the moments about the coordinate axes of the force exerted on A must be as follows: $|M_x| \leq 10$ kN · m, $|M_y| \leq 6$ kN · m, $|M_z| \leq 16$ kN · m.

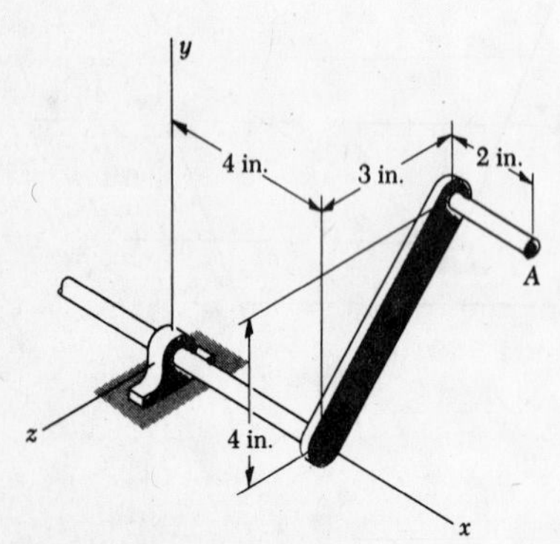

Fig. P3.34 and P3.35

3.34 The primary purpose of the crank shown is, of course, to produce a moment about the x axis. Show that a single force acting at A and having a moment M_x different from zero about the x axis must also have a moment different from zero about at least one of the other coordinate axes.

3.35 A single force **F** of unknown magnitude and direction acts at point A of the crank shown. Determine the moment M_x of **F** about the x axis, knowing that $M_y = +180$ lb·in. and $M_z = -320$ lb·in.

3.36 A force **P** of magnitude 25 lb acts on a bent rod as shown. Determine the moment of **P** about (*a*) a line joining points *C* and *F*, (*b*) a line joining points *O* and *C*.

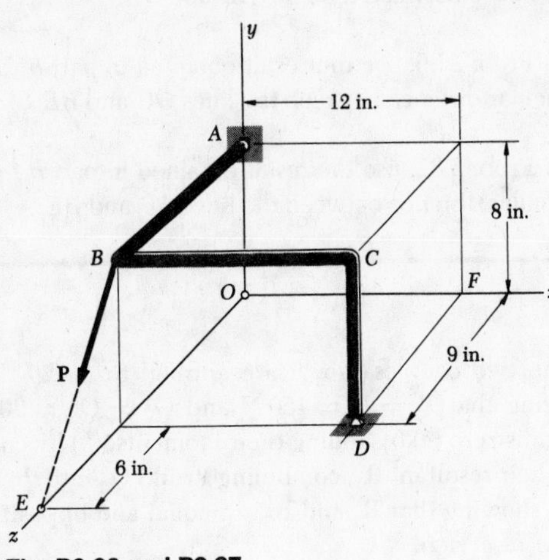

Fig. P3.36 and P3.37

3.37 A force **P** of magnitude 25 lb acts on a bent rod as shown. Determine the moment of **P** about (*a*) a line joining points *A* and *C*, (*b*) a line joining points *A* and *D*.

3.38 Two rods are welded together to form a T-shaped lever which is acted upon by a 650-N force as shown. Determine the moment of the force about rod *AB*.

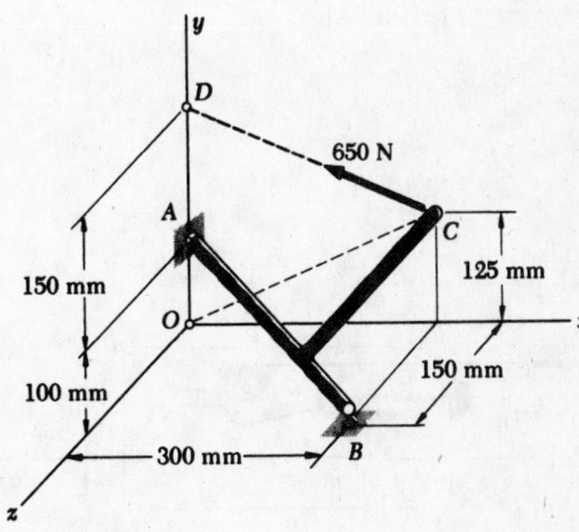

Fig. P3.38

3.39 The rectangular plate *ABCD* is held by hinges along its edge *AD* and by the wire *BE*. Knowing that the tension in the wire is 546 N, determine the moment about *AD* of the force exerted by the wire at point *B*.

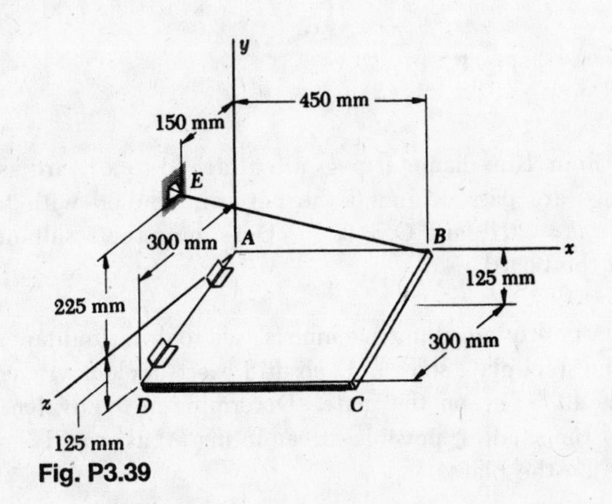

Fig. P3.39

***3.40** Use the result obtained in Prob. 3.38 to determine the perpendicular distance between the lines *AB* and *CD*.

***3.41** Use the result obtained in Prob. 3.39 to determine the perpendicular distance between the lines *AD* and *BE*.

***3.42** In Prob. 3.36, use the result obtained in part *b* to determine the perpendicular distance between the lines *OC* and *BE*.

***3.43** In Prob. 3.37, use the result obtained in part *b* to determine the perpendicular distance between the lines *AD* and *BE*.

SECTIONS 3.12 to 3.16

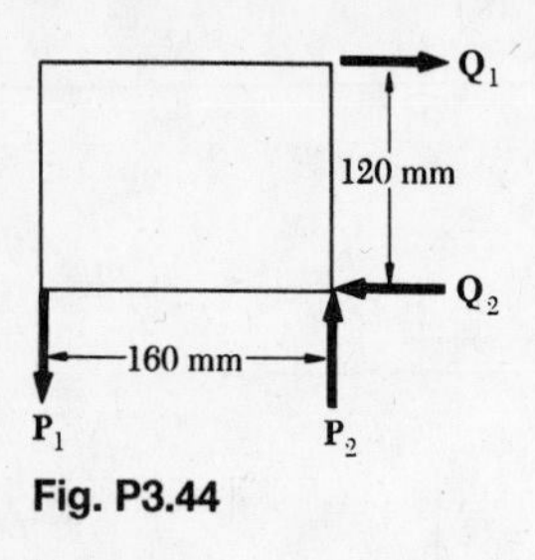

Fig. P3.44

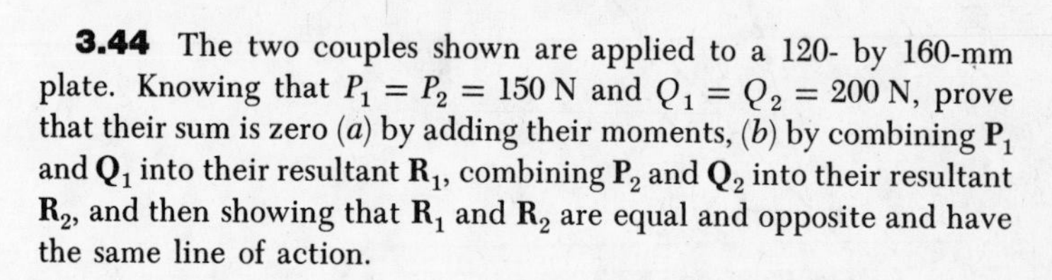

3.44 The two couples shown are applied to a 120- by 160-mm plate. Knowing that $P_1 = P_2 = 150$ N and $Q_1 = Q_2 = 200$ N, prove that their sum is zero (*a*) by adding their moments, (*b*) by combining $\mathbf{P}_1$ and $\mathbf{Q}_1$ into their resultant $\mathbf{R}_1$, combining $\mathbf{P}_2$ and $\mathbf{Q}_2$ into their resultant $\mathbf{R}_2$, and then showing that $\mathbf{R}_1$ and $\mathbf{R}_2$ are equal and opposite and have the same line of action.

3.45 A couple formed by two 975-N forces is applied to the pulley assembly shown. Determine an equivalent couple which is formed by (*a*) vertical forces acting at *A* and *C*, (*b*) the smallest possible forces acting at *B* and *D*, (*c*) the smallest possible forces which can be attached to the assembly.

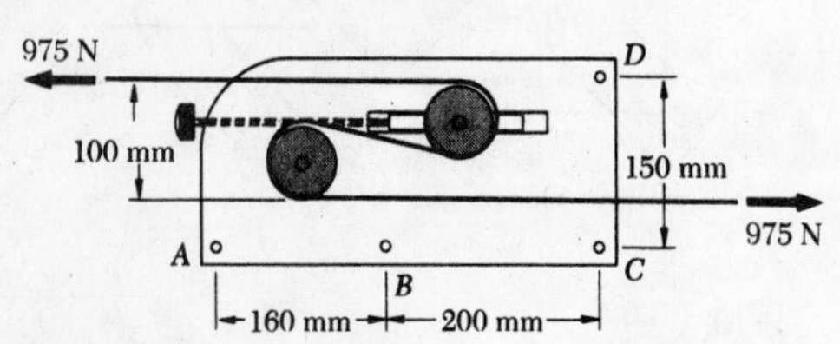

Fig. P3.45

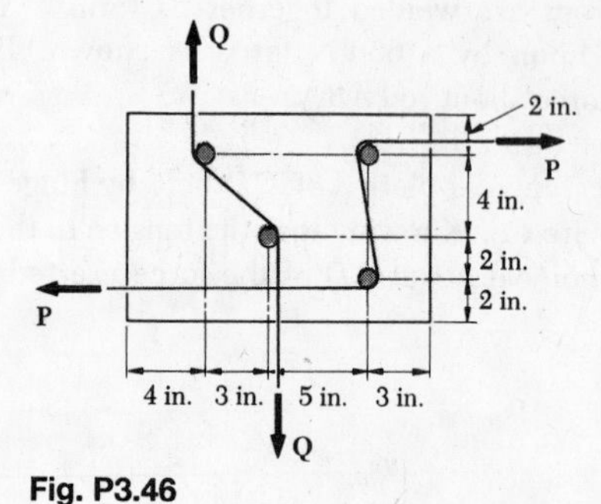

Fig. P3.46

3.46 Four 1-in.-diameter pegs are attached to a board as shown. Two strings are passed around the pegs and pulled with forces of magnitude $P = 20$ lb and $Q = 35$ lb. Determine the resultant couple acting on the board.

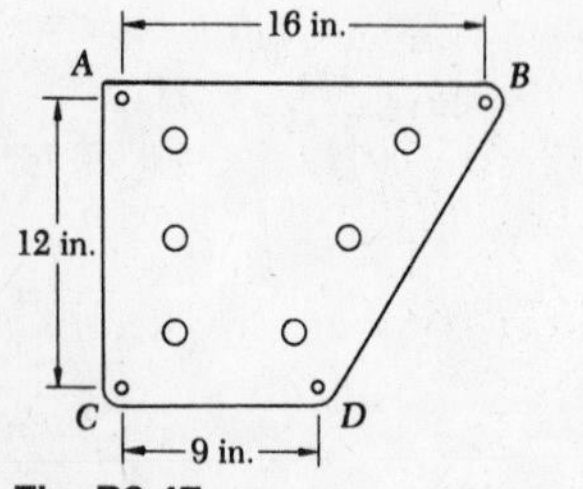

Fig. P3.47

3.47 A multiple-drilling machine is used to drill simultaneously six holes in the steel plate shown. Each drill exerts a clockwise couple of magnitude 40 lb · in. on the plate. Determine an equivalent couple formed by the smallest possible forces acting (*a*) at *A* and *C*, (*b*) at *A* and *D*, (*c*) on the plate.

3.48 The axles and drive shaft of an automobile are acted upon by the three couples shown. Replace these three couples by a single equivalent couple.

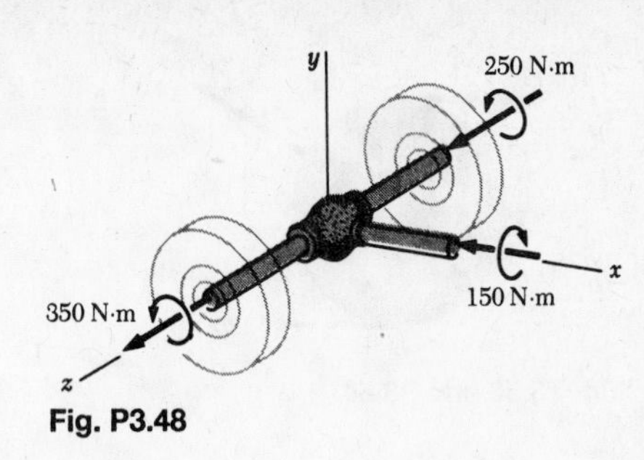

Fig. P3.48

3.49 The gearbox is acted upon by the three couples shown. Replace these three couples by a single equivalent couple.

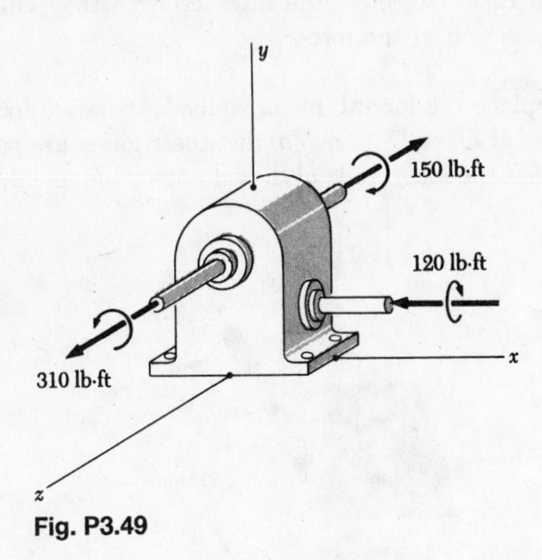

Fig. P3.49

3.50 The couple vectors $\mathbf{M}_1$ and $\mathbf{M}_2$ represent couples which are contained in the planes ABC and ACD, respectively. Assuming that $M_1 = M_2 = M$, determine a single couple equivalent to the two given couples.

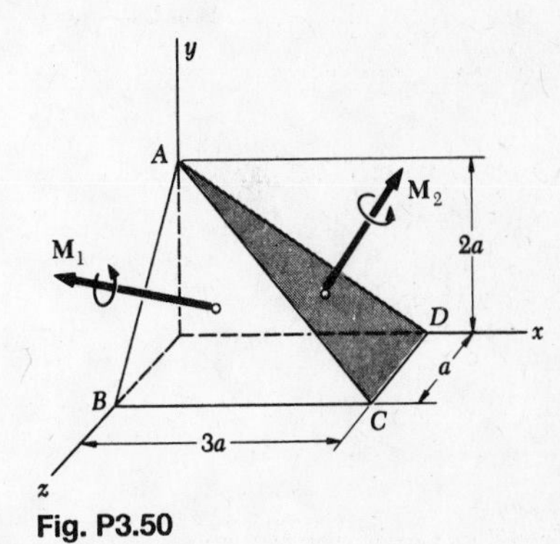

Fig. P3.50

3.51 Three shafts are connected to a gearbox as shown. Shaft A is horizontal and shafts B and C lie in the vertical yz plane. Determine the components of the resultant couple exerted on the gearbox.

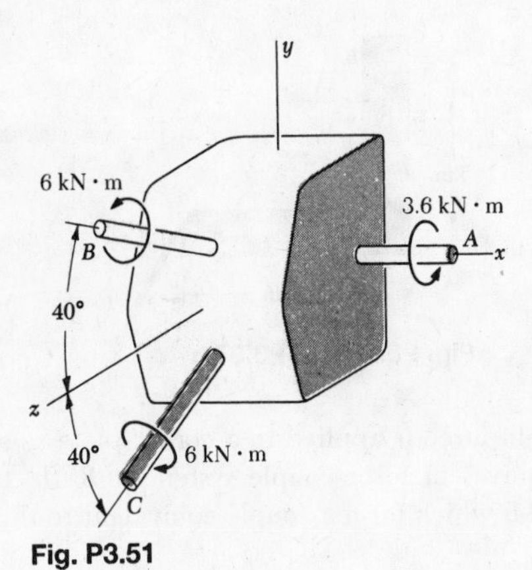

Fig. P3.51

3.52 A crane column supports a 16-kip load as shown. Reduce the load to an axial force along AB and a couple.

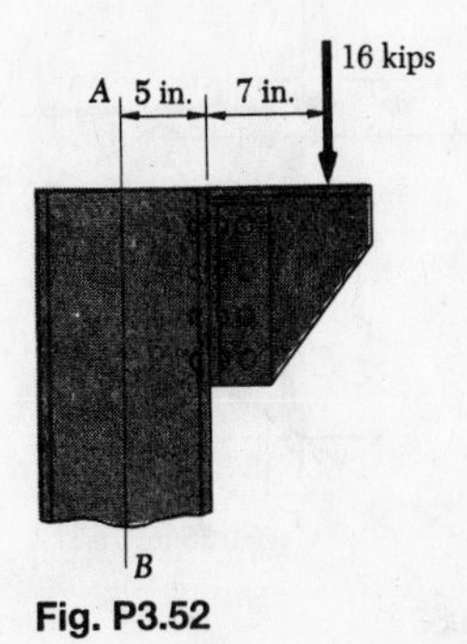

Fig. P3.52

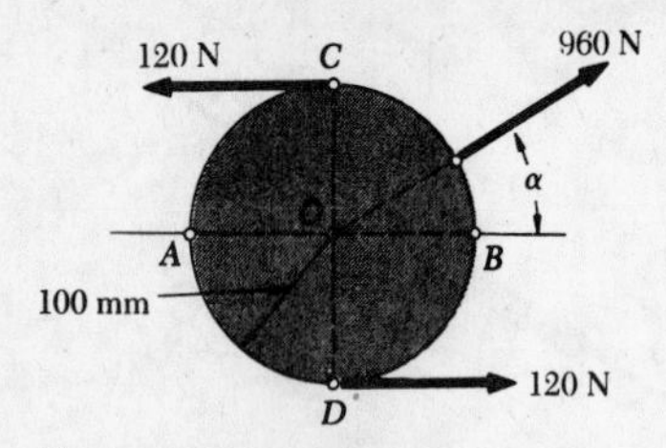

Fig. P3.53 and P3.54

3.53 The force and couple shown are to be replaced by an equivalent single force. Determine the required value of α so that the line of action of the single equivalent force will pass through point B.

3.54 Knowing that $\alpha = 60°$, replace the force and couple shown by a single force applied at a point located (*a*) on line *AB*, (*b*) on line *CD*. In each case determine the distance from the center O to the point of application of the force.

3.55 Replace the force **P** by an equivalent system formed by two parallel forces at *B* and *C*. Show (*a*) that these forces are parallel to the force **P**, (*b*) that the magnitude of these forces is independent of both α and β.

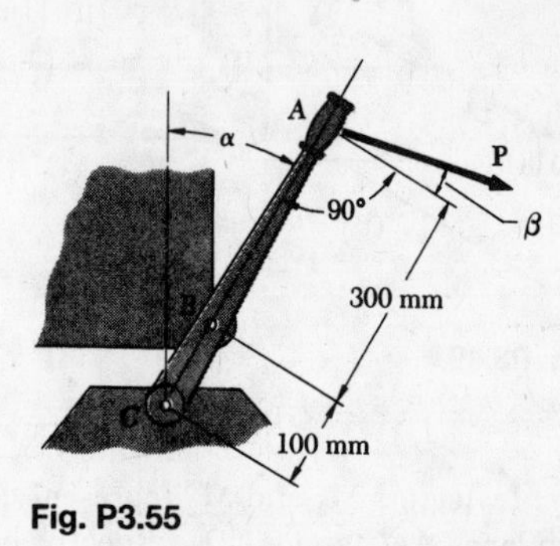

Fig. P3.55

3.56 A 50-lb force is applied to a corner plate as shown. Determine (*a*) an equivalent force-couple system at *A*, (*b*) two horizontal forces at *A* and *B* which form a couple equivalent to the couple found in part *a*.

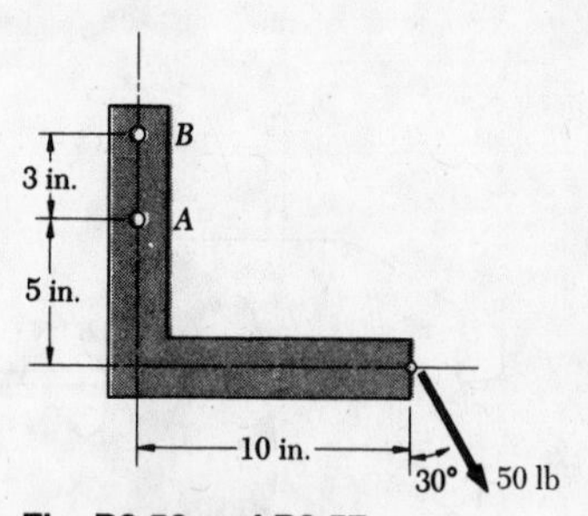

Fig. P3.56 and P3.57

3.57 A 50-lb force is applied to a corner plate as shown. Determine (*a*) an equivalent force-couple system at *B*, (*b*) two horizontal forces at *A* and *B* which form a couple equivalent to the couple found in part *a*.

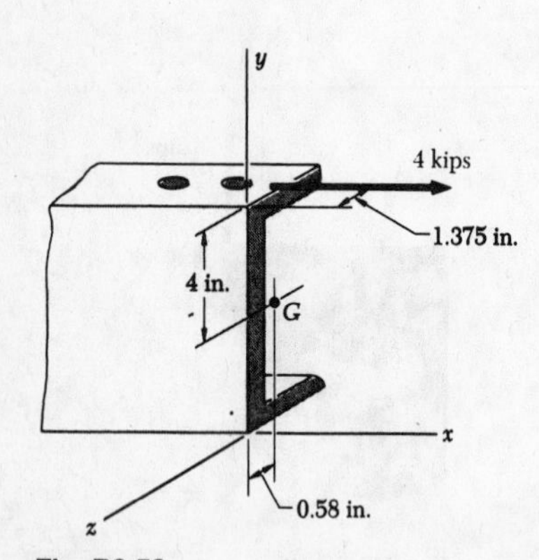

Fig. P3.58

3.58 A 4-kip force is applied on the outside face of the flange of a steel channel. Determine the components of the force and couple at G which are equivalent to the 4-kip load.

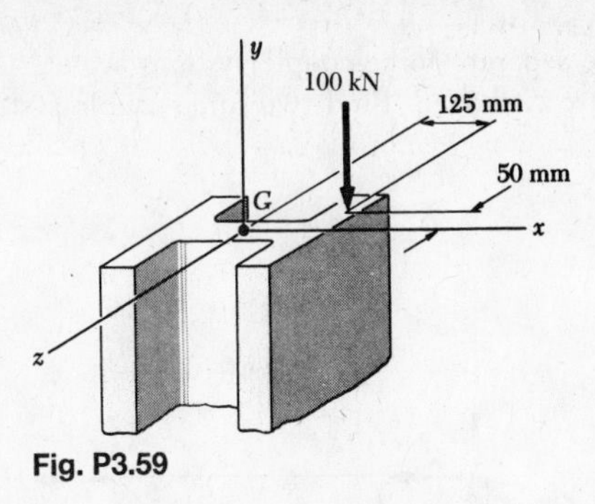

Fig. P3.59

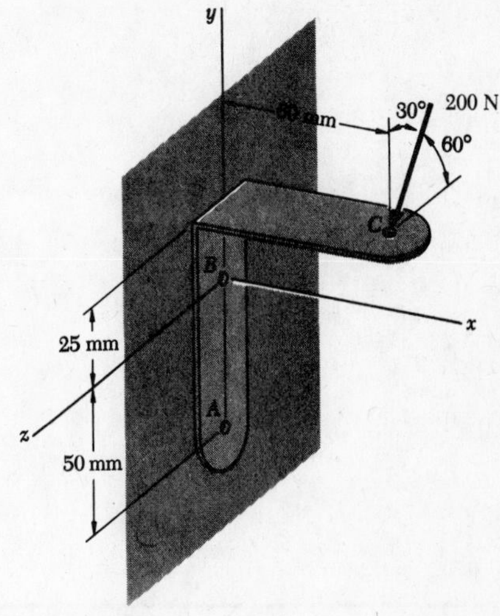

Fig. P3.60

3.59 A 100-kN load is applied eccentrically on a column. Determine the components of the force and couple at G which are equivalent to the 100-kN load.

3.60 A 200-N force is applied as shown on the bracket ABC. Determine the components of the force and couple at A which are equivalent to this force.

3.61 A precast-concrete wall section is temporarily held by cables as shown. The tension in cable AB is 700 lb. Replace the force exerted on the wall section at A by a force-couple system located (a) at the origin of coordinates O, (b) at point E.

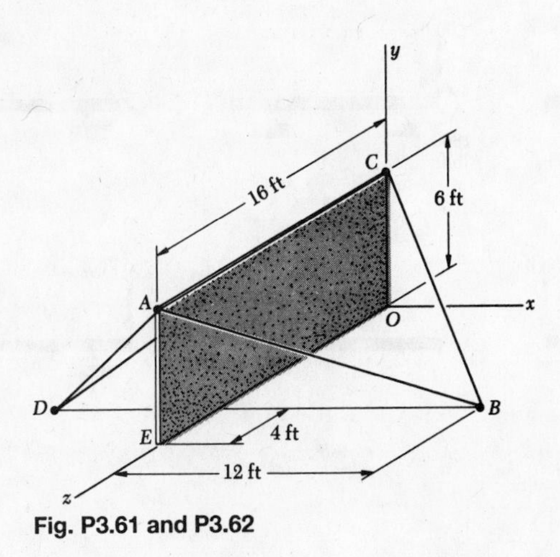

Fig. P3.61 and P3.62

3.62 A precast-concrete wall section is temporarily held by cables as shown. The tension in cable BC is 900 lb. Replace the force exerted on the wall section at C by a force-couple system located (a) at the origin of coordinates O, (b) at point E.

3.63 Five separate force-couple systems act at the corners of a rectangular box as shown. Find two force-couple systems which are equivalent.

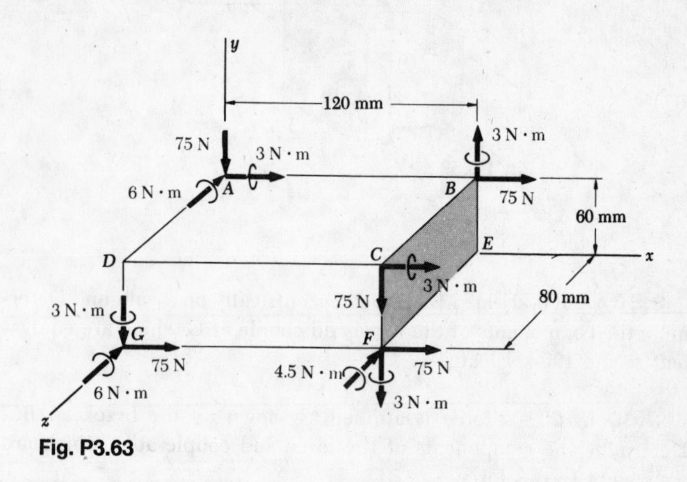

Fig. P3.63

SECTIONS 3.17 to 3.21

3.64 A 12-ft beam is loaded in the various ways represented in the figure. Find two loadings which are equivalent.

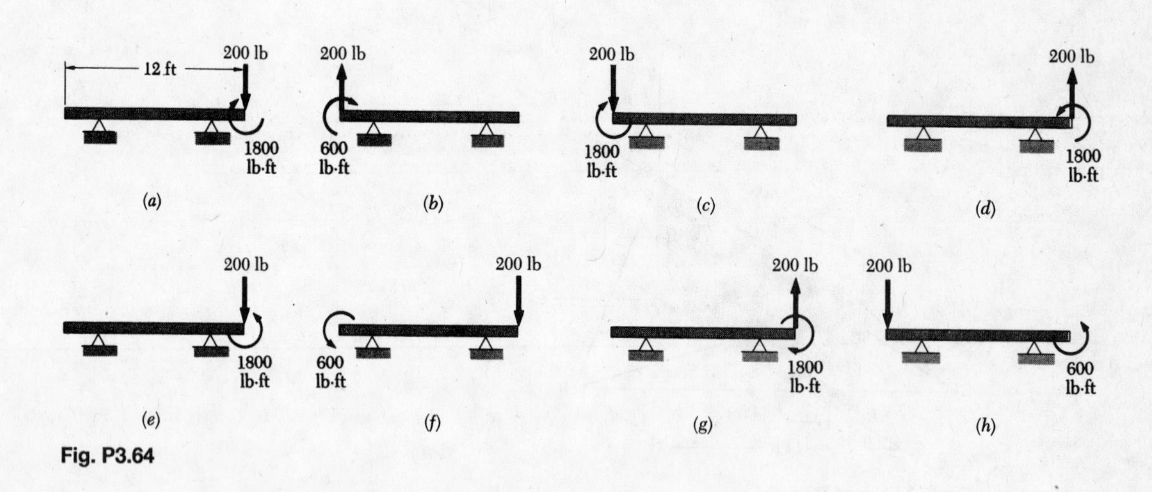

Fig. P3.64

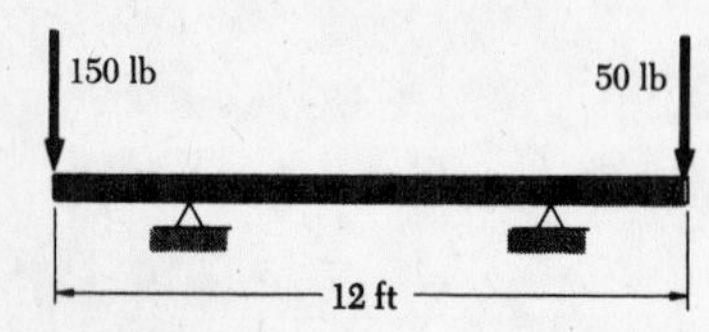

Fig. P3.65

3.65 A 12-ft beam is loaded as shown. Determine the loading of Prob. 3.64 which is equivalent to this loading.

3.66 Determine the distance from point A to the line of action of the resultant of the three forces shown when (a) $a = 1$ m, (b) $a = 1.5$ m, (c) $a = 2.5$ m.

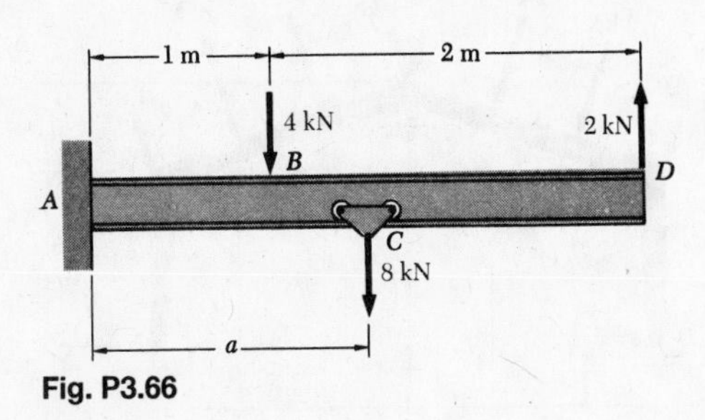

Fig. P3.66

3.67 Two parallel forces **P** and **Q** are applied at the ends of a beam AB of length L. Find the distance x from A to the line of action of their resultant. Check the formula obtained by assuming $L = 200$ mm and (a) $P = 50$ N down, $Q = 150$ N down; (b) $P = 50$ N down, $Q = 150$ N up.

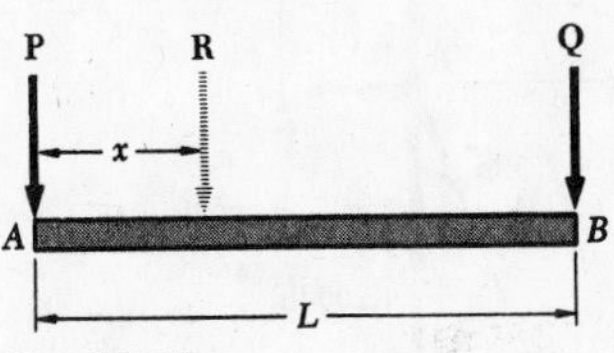

Fig. P3.67

3.68 For the truss and loading shown, determine the resultant of the loads and the distance from point A to its line of action.

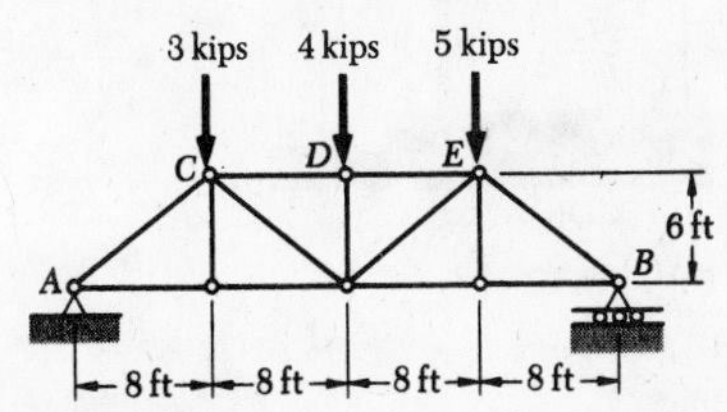

Fig. P3.68

3.69 Four packages are transported at constant speed from A to B by the conveyor. At the instant shown, determine the resultant of the loading and the location of its line of action.

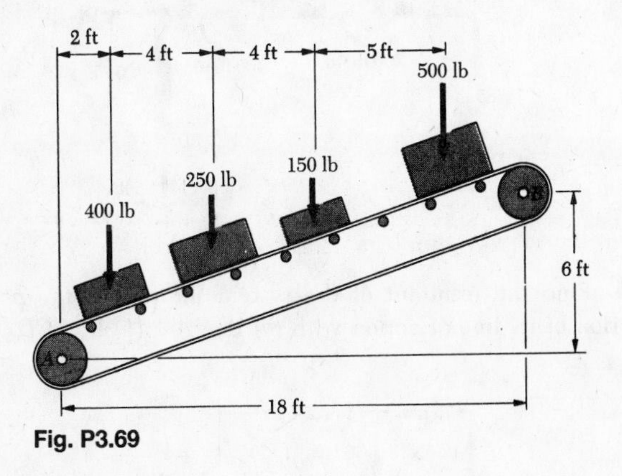

Fig. P3.69

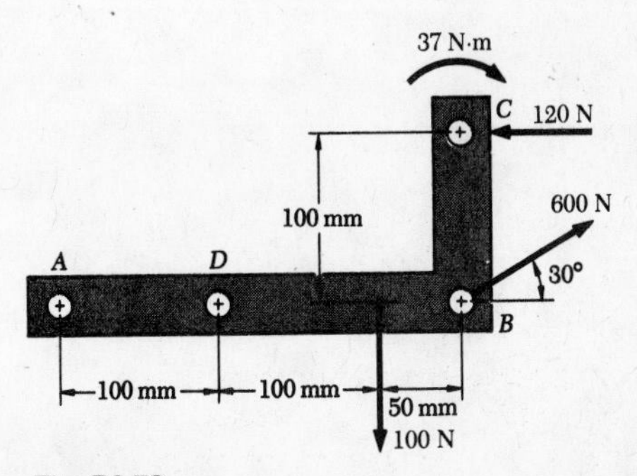

3.70 An angle bracket is subjected to the system of forces shown. Find the resultant of the system and the point of intersection of its line of action with (a) line AB, (b) line BC.

Fig. P3.70

3.71 The roof of a building frame is subjected to the wind loading shown. Determine (*a*) the equivalent force-couple system at *D*, (*b*) the resultant of the loading and its line of action.

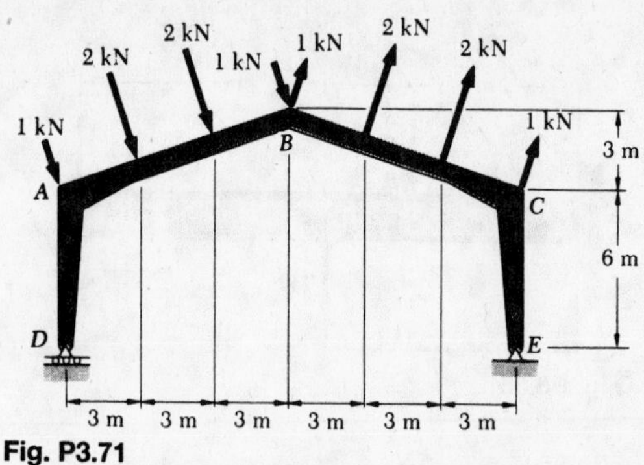

Fig. P3.71

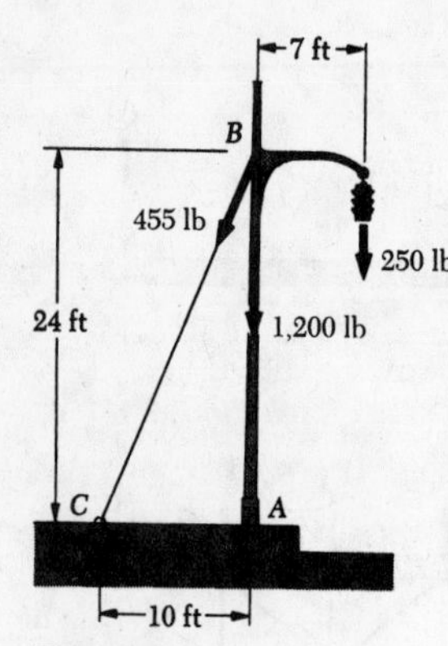

Fig. P3.72

3.72 Three forces act as shown on a traffic-signal pole. Determine (*a*) the equivalent force-couple system at *A*, (*b*) the resultant of the system and the point of intersection of its line of action with the pole.

3.73 A bracket is subjected to the system of forces and couples shown. Find the resultant of the system and the point of intersection of its line of action with (*a*) line *AB*, (*b*) line *BC*, (*c*) line *CD*.

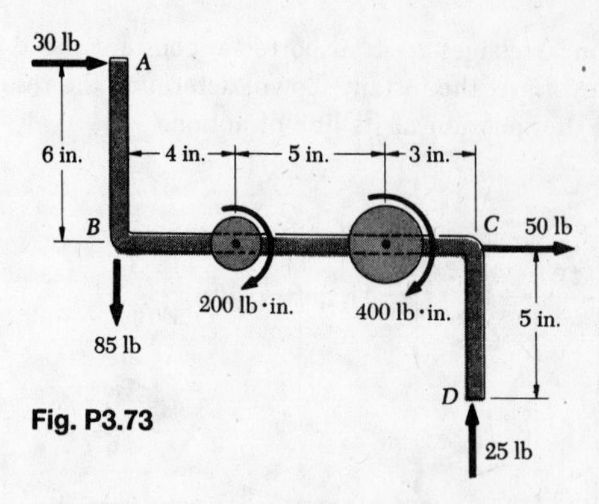

Fig. P3.73

3.74 Find the resultant of the system shown and the point of intersection of its line of action with (*a*) line *AC*, (*b*) line *CD*.

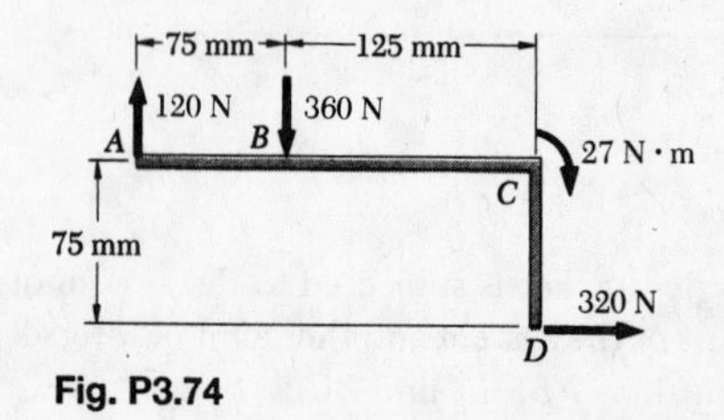

Fig. P3.74

3.75 In order to move a 70.6-kg crate, two men push on it while two other men pull on it by means of ropes. The force exerted by man A is 600 N and that exerted by man B is 200 N; both forces are horizontal. Man C pulls with a force equal to 320 N and man D with a force equal to 480 N. Both cables form an angle of 30° with the vertical. Determine the resultant of all forces acting on the crate.

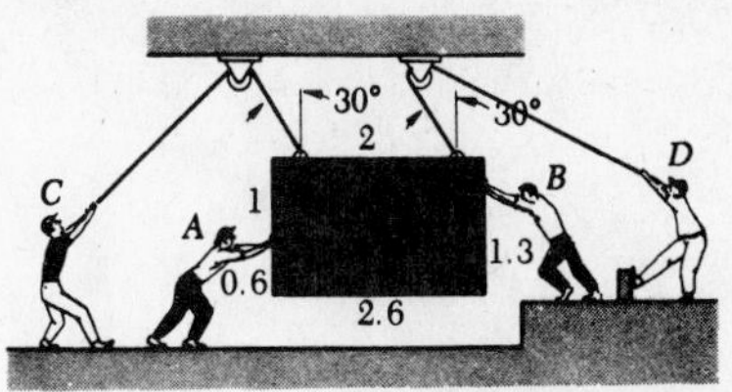
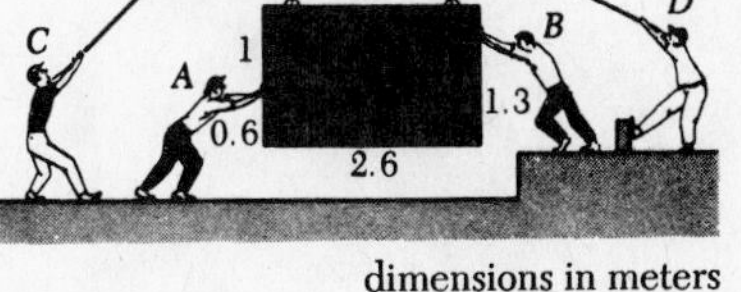

Fig. P3.75

3.76 Two forces are applied to the vertical post as shown. Determine the force and couple at O equivalent to the two forces.

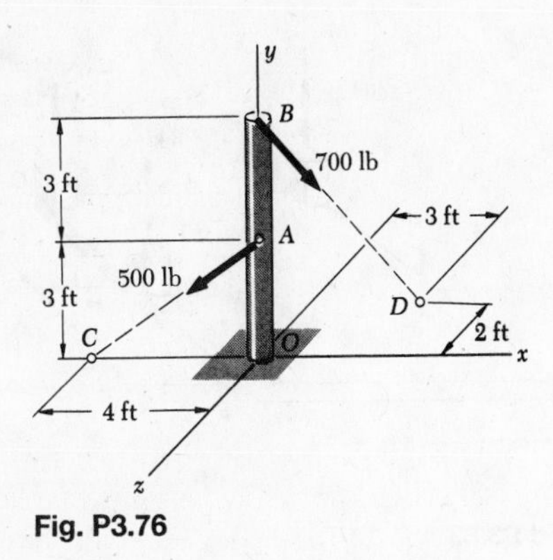

Fig. P3.76

Fig. P3.77

3.77 In drilling a hole in a wall, a man applies a vertical 30-lb force at B on the brace and bit, while pushing at C with a 10-lb force. The brace lies in the horizontal xz plane. (a) Determine the other components of the total force which should be exerted at C if the bit is not to be bent about the y and z axes (i.e., if the system of forces applied on the brace is to have zero moment about both the y and z axes). (b) Reduce the 30-lb force and the total force at C to an equivalent force and couple at A.

3.78 In order to tighten the joint between the tapped faucet A and the pipe AC, a plumber uses two pipe wrenches as shown. By exerting a 250-N force on each wrench, at a distance of 200 mm from the axis of the pipe and in a direction perpendicular to the pipe and to the wrench, he prevents the pipe from rotating, and thus avoids loosening or further tightening the joint between the pipe and the tapped elbow C. Replace the two given forces by an equivalent force-couple system at D and determine whether the plumber's action tends to tighten or loosen the joint between (a) pipe CD and elbow D, (b) elbow D and pipe DE. Assume all threads to be right-handed.

3.79 In Prob. 3.78, replace the two given forces by an equivalent force-couple system at E and determine whether the plumber's action tends to tighten or loosen the joint between (a) pipe DE and elbow E, (b) elbow E and pipe EO. Assume all threads to be right-handed.

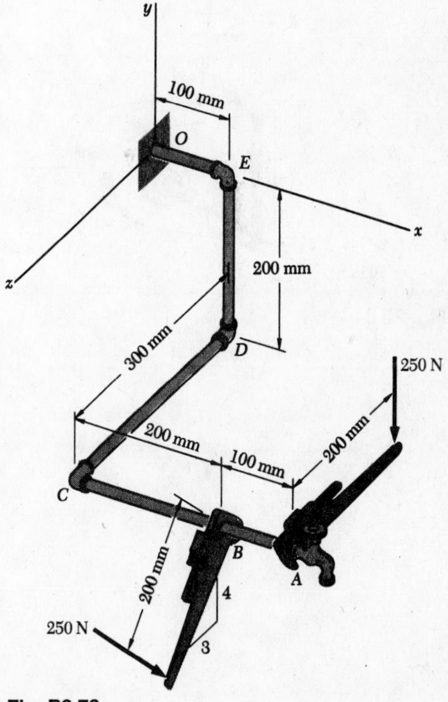

Fig. P3.78

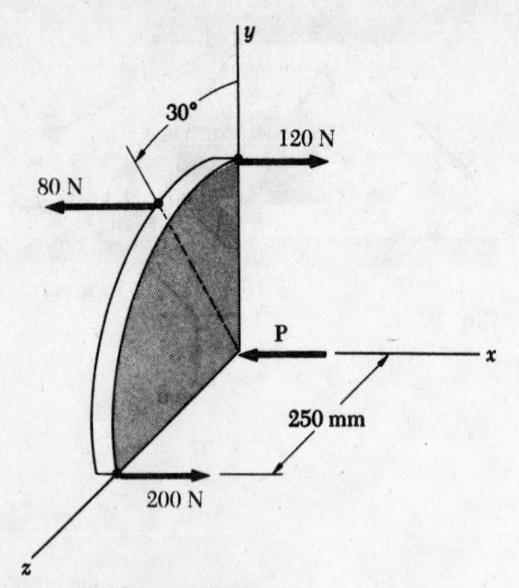

Fig. P3.80 and P3.81

3.80 Four horizontal forces act on a vertical quarter-circular plate of radius 250 mm. Determine the magnitude and point of application of the resultant of the four forces if $P = 40$ N.

3.81 Determine the magnitude of the force **P** for which the resultant of the four forces acts on the rim of the plate.

3.82 Four column loads act on a square foundation mat as shown. Determine the magnitude and point of application of the resultant of the four loads.

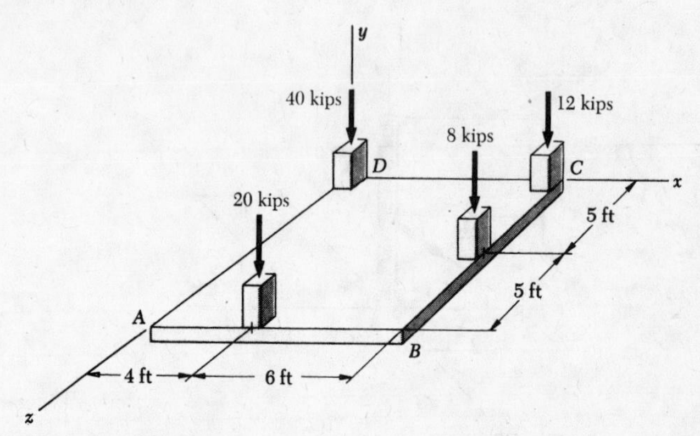

Fig. P3.82 and P3.83

3.83 Four column loads act on a square foundation mat as shown. Determine the additional vertical loads at A and B if the resultant of the six loads is to pass through the center of the mat.

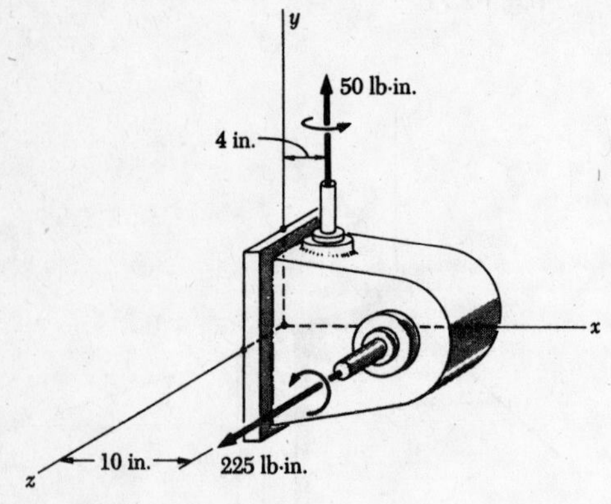

Fig. P3.84

***3.84** The worm-gear speed reducer shown weighs 75 lb; the center of gravity is located on the x axis at $x = 8$ in. Replace the weight and couples shown by a wrench. (Specify the axis and pitch of the wrench.)

***3.85** Replace the two forces shown by (a) a force-couple system at the origin, (b) a wrench. (Specify the axis and pitch of the wrench.)

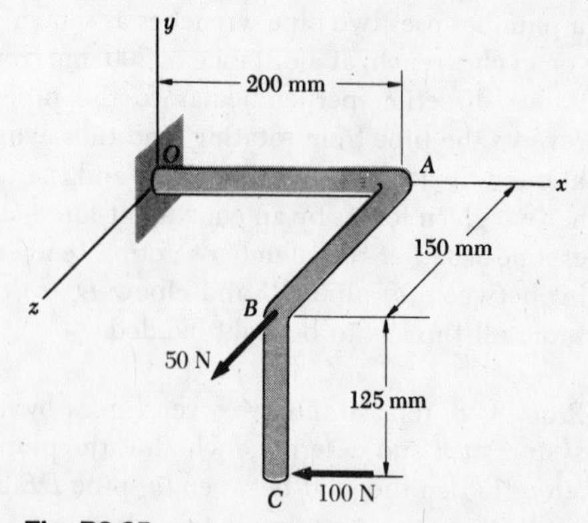

Fig. P3.85

***3.86** Two forces of magnitude P act along the diagonals of the faces of a cube of side a as shown. Replace the two forces by a system consisting of (*a*) a single force at O and a couple, (*b*) a wrench. (Specify the axis and pitch of the wrench.)

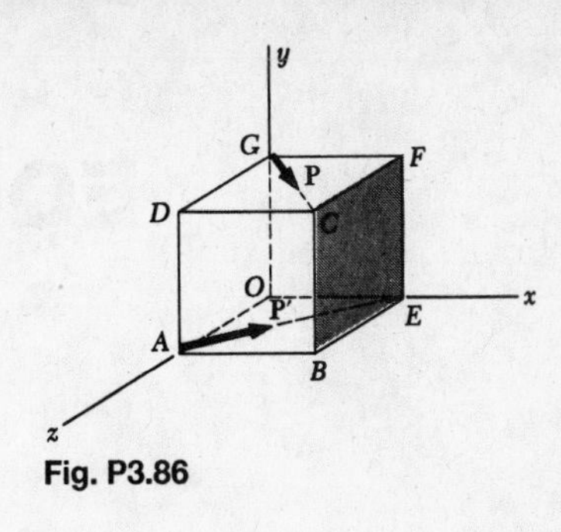

Fig. P3.86

***3.87** Knowing that $R = 70$ lb and $M = 140$ lb · in., replace the given wrench by a system of two forces chosen in such a way that one force acts at point B and the other force lies in the xz plane.

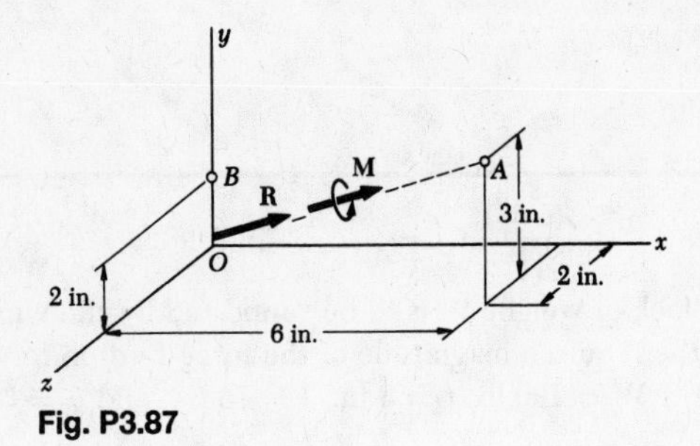

Fig. P3.87

CHAPTER 4
EQUILIBRIUM OF RIGID BODIES

SECTIONS 4.1 to 4.5

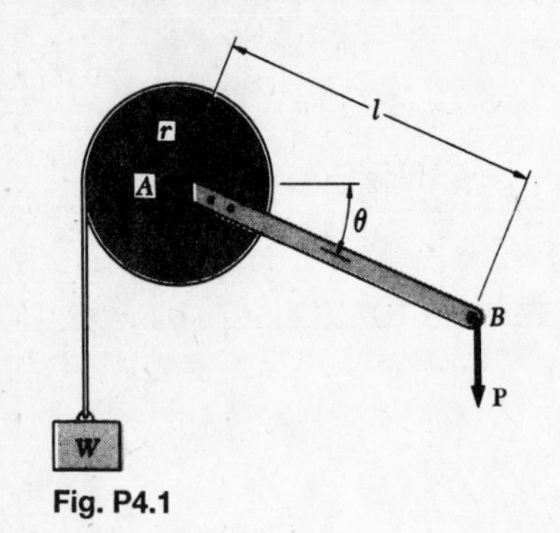

Fig. P4.1

4.1 A block of weight W is to be supported by the winch shown. Determine the required magnitude of the force $\mathbf{P}$ (*a*) in terms of W, r, l, and θ, (*b*) if $W = 100$ lb, $r = 3$ in., $l = 15$ in., and $\theta = 60°$.

4.2 Two external shafts of a gearbox carry torques as shown. Determine the vertical components of the forces which must be exerted by the bolts at A and B to maintain the gearbox in equilibrium.

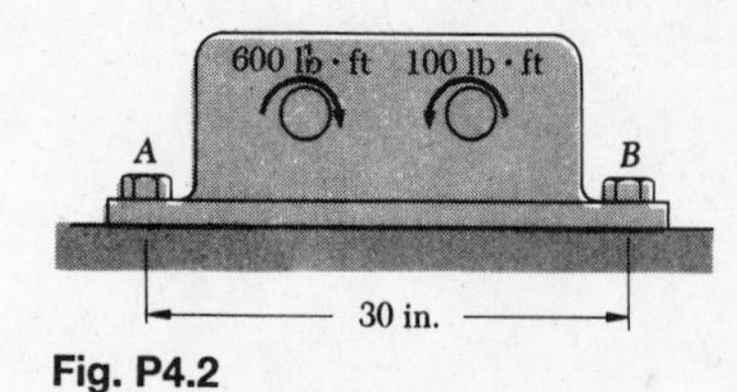

Fig. P4.2

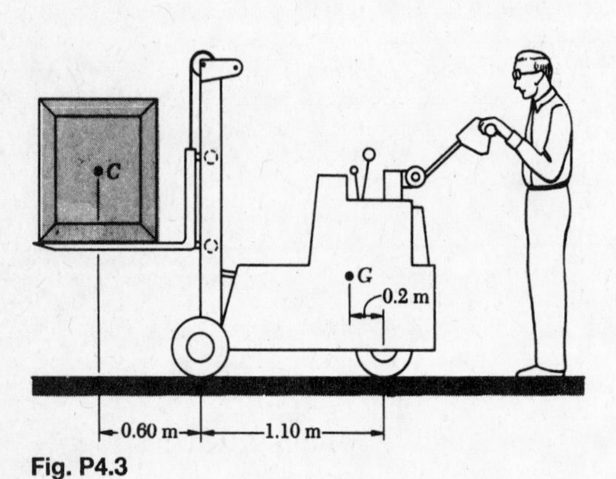

Fig. P4.3

4.3 The 600-kg forklift truck is used to hold the 150-kg crate C in the position shown. Determine the reactions (*a*) at each of the two wheels A (one wheel on each side of the truck), (*b*) at the single steerable wheel B.

4.4 The maximum allowable value for each of the reactions is 150 kN and the reaction at A must be directed upward. Neglecting the weight of the beam, determine the range of values of P for which the beam is safe.

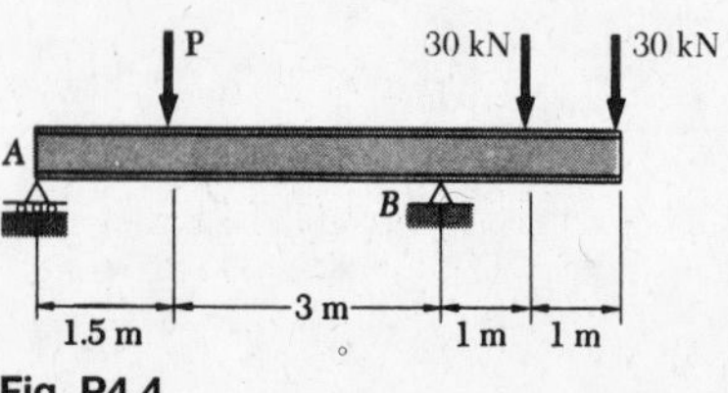

Fig. P4.4

4.5 The 10-ft beam AB rests upon, but is not attached to, supports at C and D. Neglecting the weight of the beam, determine the range of values of P for which the beam will remain in equilibrium.

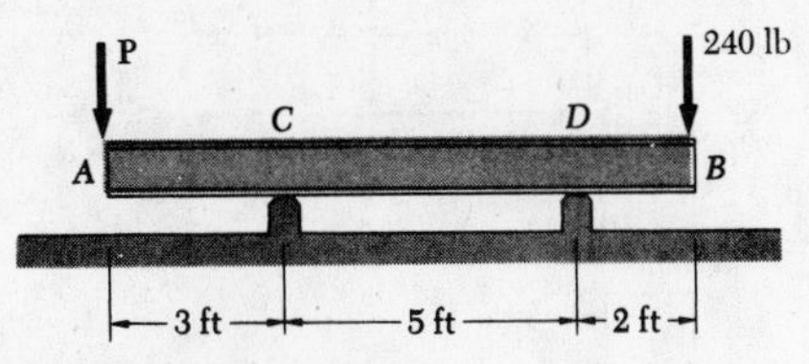

Fig. P4.5 and P4.6

4.6 The 10-ft uniform beam AB weighs 100 lb; it rests upon, but is not attached to, supports at C and D. Determine the range of values of P for which the beam will remain in equilibrium.

4.7 The 40-ft boom AB weighs 2 kips; the distance from the axle A to the center of gravity G of the boom is 20 ft. For the position shown, determine the tension T in the cable and the reaction at A.

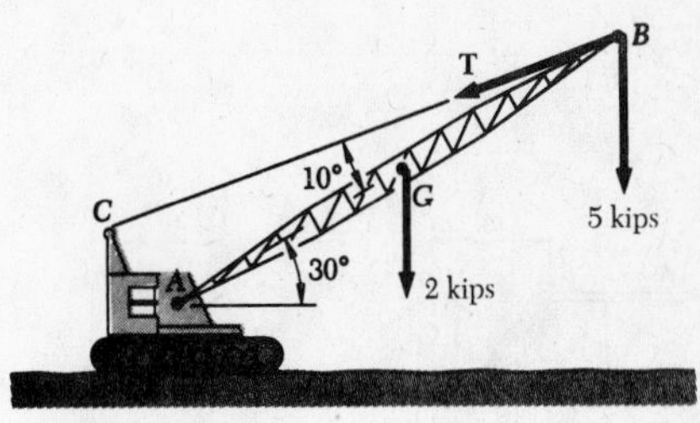

Fig. P4.7

4.8 The ladder AB, of length L and weight W, can be raised by the cable BC. Determine the tension T required to raise end B just off the floor (*a*) in terms of W and θ, (*b*) if $h = 8$ ft, $L = 10$ ft, and $W = 35$ lb.

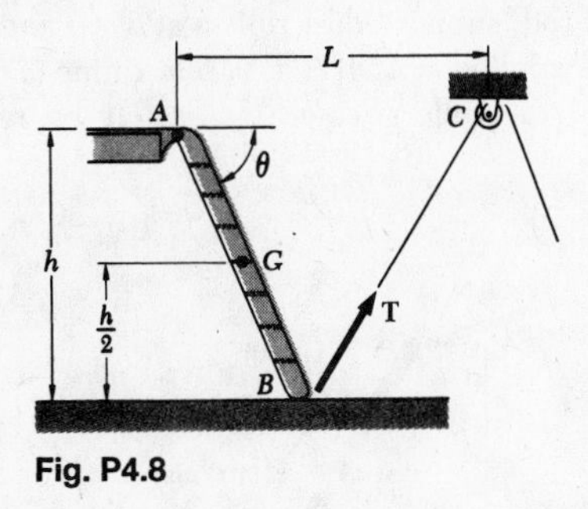

Fig. P4.8

4.9 Knowing that the magnitude of the vertical force **P** is 400 N, determine (*a*) the tension in the cable CD, (*b*) the reaction at B.

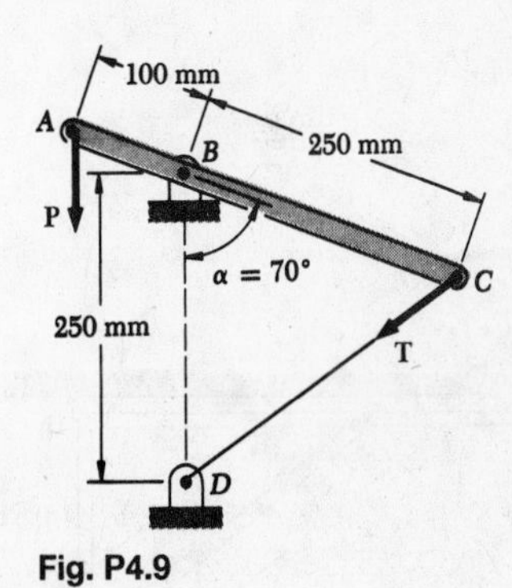

Fig. P4.9

4.10 Determine the reactions at A and B for the loading shown.

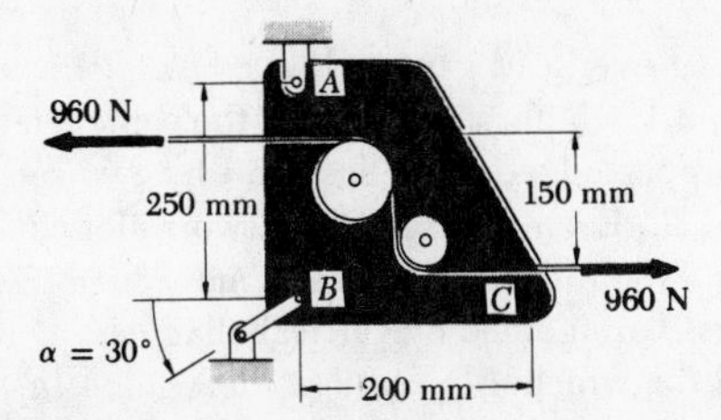

Fig. P4.10

4.11 Determine the reactions at A and B when $\alpha = 60°$.

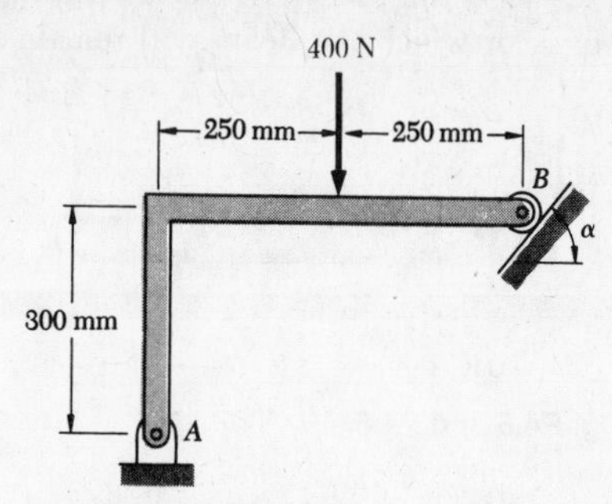

Fig. P4.11 and P4.12

4.12 Determine the reactions at A and B when $\alpha = 90°$.

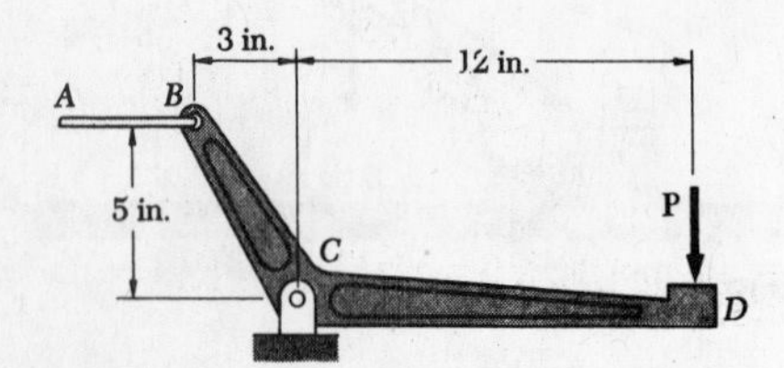

Fig. P4.13 and P4.14

4.13 The required tension in cable AB is 300 lb. Determine (*a*) the vertical force **P** which must be applied to the pedal, (*b*) the corresponding reaction at C.

4.14 Determine the maximum tension which may be developed in cable AB if the maximum allowable magnitude of the reaction at C is 650 lb.

4.15 A light rod, supported by rollers at B, C, and D, is subjected to an 800-N force applied at A. If $\beta = 0$, determine (*a*) the reactions at B, C, and D, (*b*) the rollers which may safely be removed for this loading.

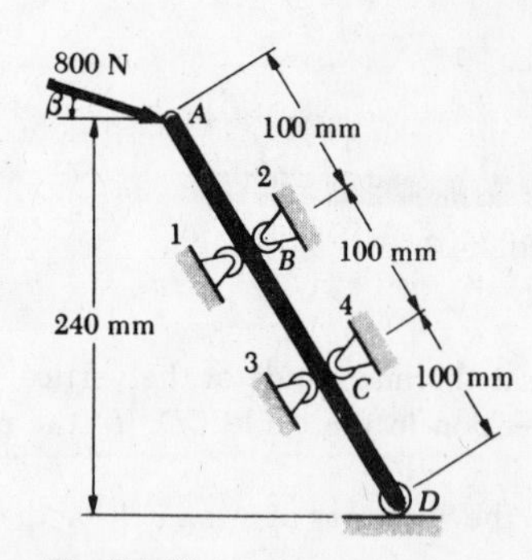

Fig. P4.15

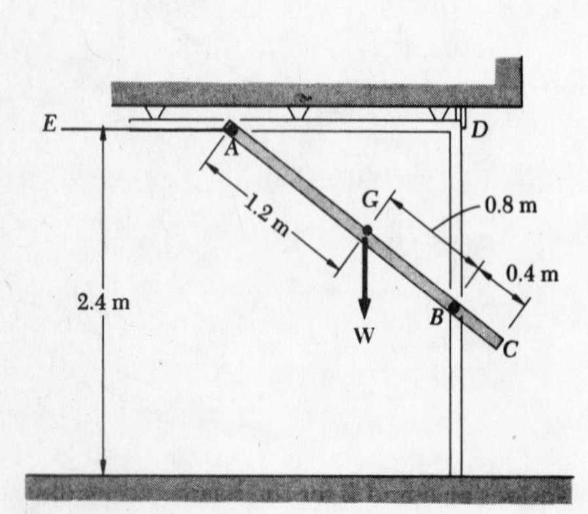

Fig. P4.16

4.16 A 90-kg overhead garage door consists of a uniform rectangular panel AC, 2.4 m high, supported by the cable AE attached at the middle of the upper edge of the door and by two sets of frictionless rollers at A and B. Each set consists of two rollers located on either side of the door. The rollers A are free to move in horizontal channels, while the rollers B are guided by vertical channels. If the door is held in the position for which $BD = 1.2$ m, determine (*a*) the tension in cable AE, (*b*) the reaction at each of the four rollers.

4.17 In Prob. 4.16, determine the minimum allowable value of the distance BD if the tension in cable AE is not to exceed 5 kN.

4.18 A movable bracket is held at rest by a cable attached at C and by frictionless rollers at A and B. For the loading shown, determine the tension in the cable and the reactions at A and B.

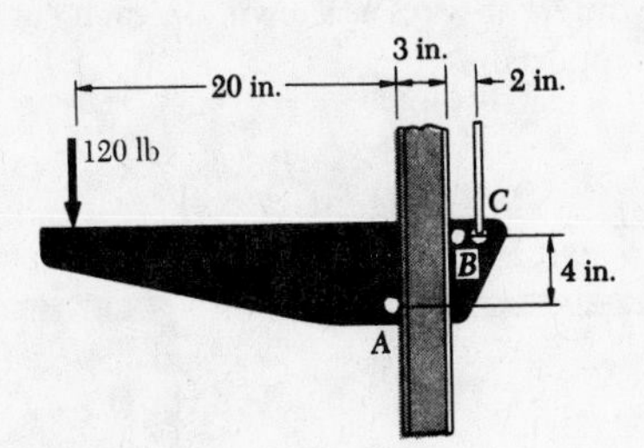

Fig. P4.18

4.19 The light bar AD is attached to collars B and C which may move freely on vertical rods. Knowing that the surface at A is smooth, determine the reactions at A, B, and C (*a*) if $\alpha = 60°$, (*b*) if $\alpha = 90°$.

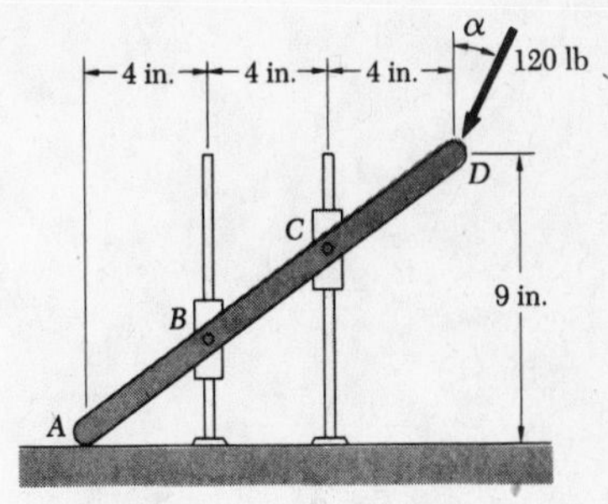

Fig. P4.19

4.20 Solve Prob. 4.19 (*a*) if $\alpha = 0°$, (*b*) if $\alpha = 30°$.

4.21 Determine the range of values of α for which the semicircular rod can be maintained in equilibrium by the small wheel at D and the rollers at B and C.

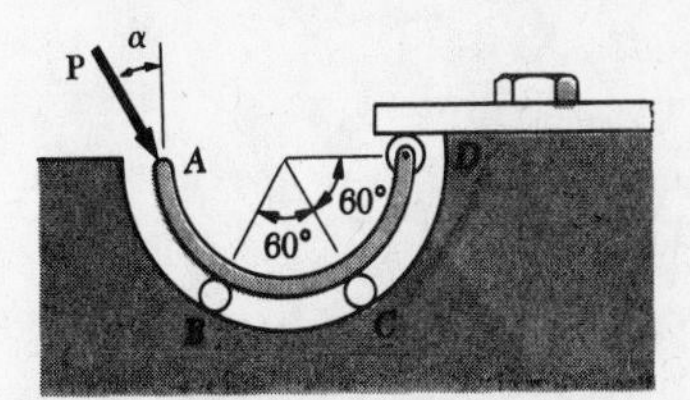

Fig. P4.21 and P4.22

4.22 Determine the reactions at B, C, and D (*a*) if $\alpha = 0$, (*b*) if $\alpha = 30°$.

4.23 A 150-kg telephone pole is used to support the ends of two wires as shown. The tension in the wire to the left is 400 N and, at the point of support, the wire forms an angle of 10° with the horizontal. (*a*) If the tension T_2 is zero, determine the reaction at the base A. (*b*) Determine the largest and smallest allowable tension T_2, if the magnitude of the couple at A may not exceed 900 N·m.

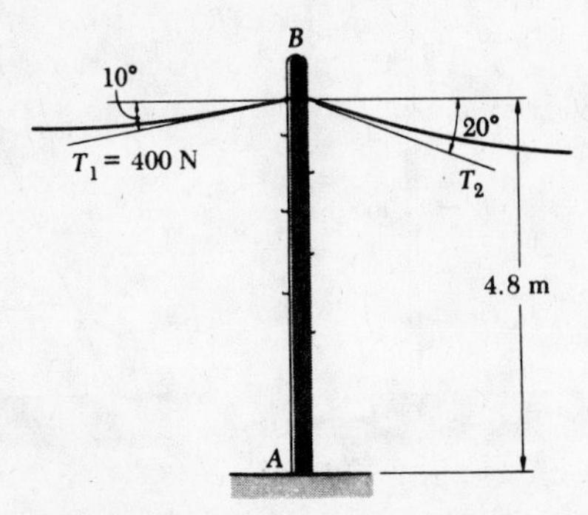

Fig. P4.23

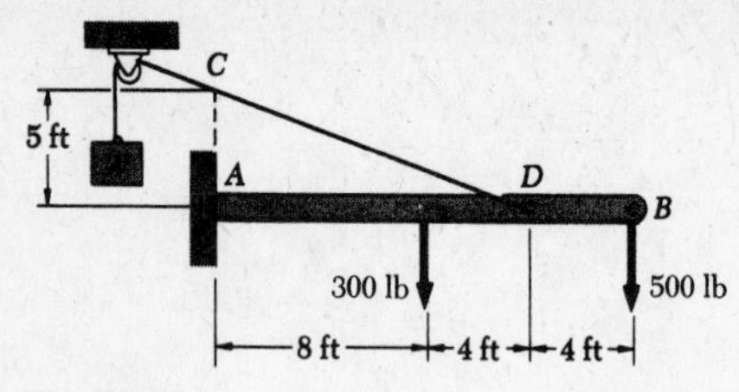

Fig. P4.24

4.24 The 300-lb beam AB carries a 500-lb load at B. The beam is held by a fixed support at A and by the cable CD which is attached to the counterweight W. (*a*) If $W = 1300$ lb, determine the reaction at A. (*b*) Determine the range of values of W for which the magnitude of the couple at A does not exceed 1500 lb·ft.

4.25 A couple **M** is applied to a bent rod AB which may be supported in four different ways as shown. In each case determine the reactions at the supports.

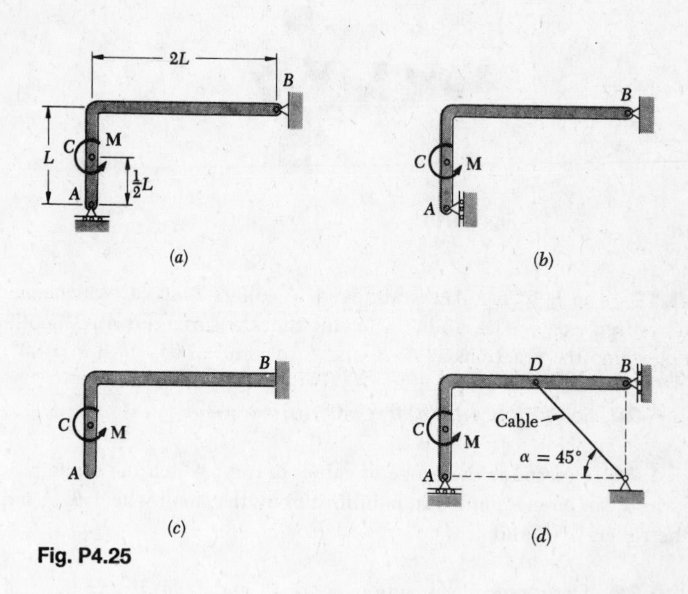

Fig. P4.25

4.26 A traffic-signal pole may be supported in the three ways shown; in part *c* the tension in cable BC is known to be 1950 N. Determine the reactions for each type of support shown.

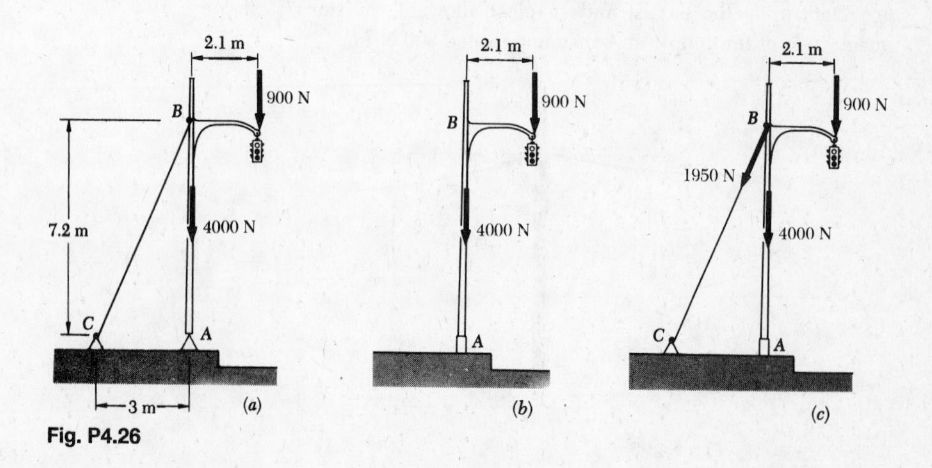

Fig. P4.26

4.27 Determine the reactions at the points of support of the bracket shown, assuming in turn the following types of connections: (*a*) a pin in a fitted hole at *A* and a pin in a vertical slot at *B*, (*b*) a pin in a horizontal slot at *A* and a pin in a fitted hole at *B*, (*c*) a firmly clinched rivet at *B*.

4.28 Determine the reactions at the points of support of the bracket shown, assuming in turn the following types of connections: (*a*) a pin in a fitted hole at *A* and a pin in a horizontal slot at *B*, (*b*) a pin in a vertical slot at *A* and a pin in a fitted hole at *B*, (*c*) a firmly clinched rivet at *A*.

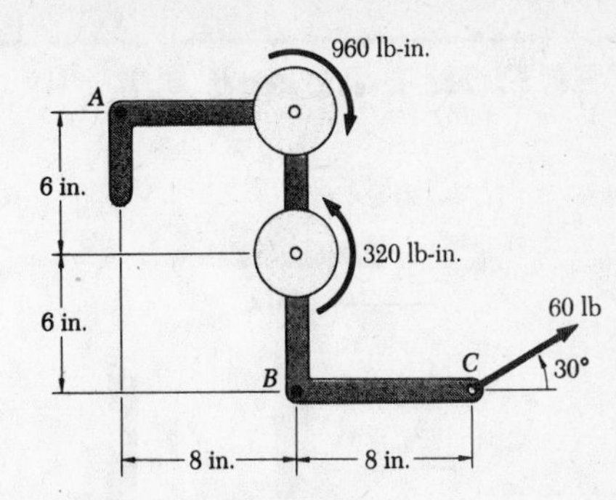

Fig. P4.27 and P4.28

4.29 The horizontal member *ABC* of the rig shown weighs 1000 lb and is supported by a pin *B* and a cable *EADC*. Since the cable passes over pulleys at *A* and *D*, the tension may be assumed to be the same in all portions of the cable. If the rig raises a 3000-lb load at a distance a = 12 ft from the vertical member *DF*, determine (*a*) the tension in the cable, (*b*) the horizontal and vertical components of the reaction at *B*.

4.30 For the rig of Prob. 4.29, determine (*a*) the maximum distance a from the vertical member *DF* at which a 3000-lb load may be supported if the maximum allowable tension in cable *EADC* is 8000 lb, (*b*) the corresponding values of the horizontal and vertical components of the reaction at *B*.

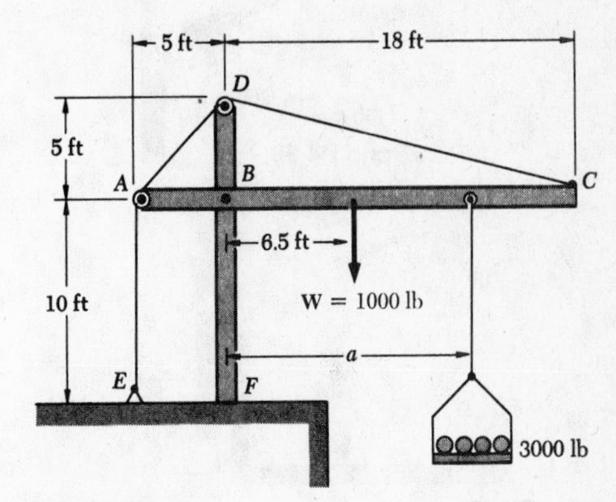

Fig. P4.29

4.31 In the pivoted motor mount, or Rockwood drive, the weight of the motor is used to maintain tension in the drive belt. When the motor is at rest, the tensions T_1 and T_2 may be assumed equal. The mass of the motor is 90 kg, and the diameter of the drive pulley is 150 mm. Assuming that the mass of the platform *AB* is negligible, determine (*a*) the tension in the belt, (*b*) the reaction at *C* when the motor is at rest.

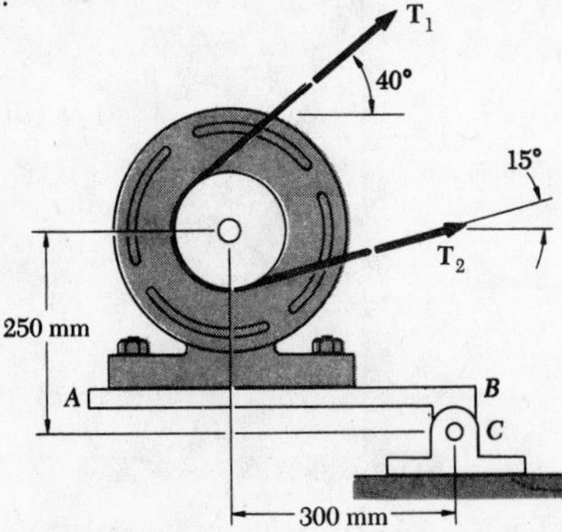

Fig. P4.31

***4.32** A uniform, slender rod of length *L* and weight *W* is held in the position shown by the horizontal force **P**. Neglecting the effect of friction at *A* and *B*, determine the angle θ corresponding to equilibrium (*a*) in terms of *P*, *W*, *L*, and β, (*b*) if *P* = 50 N, *W* = 100 N, *L* = 750 mm, and β = 60°.

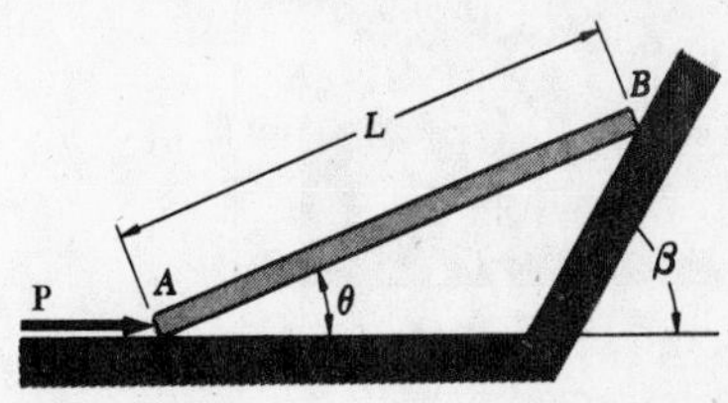

Fig. P4.32

SECTIONS 4.6 and 4.7

4.33 Using the method of Sec. 4.7, solve Prob. 4.11.

4.34 Using the method of Sec. 4.7, solve Prob. 4.9.

4.35 Using the method of Sec. 4.7, solve Prob. 4.13.

4.36 Using the method of Sec. 4.7, solve Prob. 4.14.

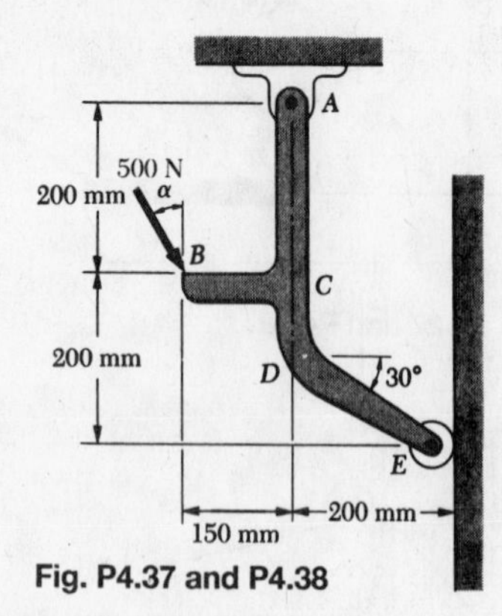

Fig. P4.37 and P4.38

4.37 Determine the reactions at A and E when $\alpha = 0$.

4.38 Determine (a) the value of α for which the reaction at A is vertical, (b) the corresponding reactions at A and E.

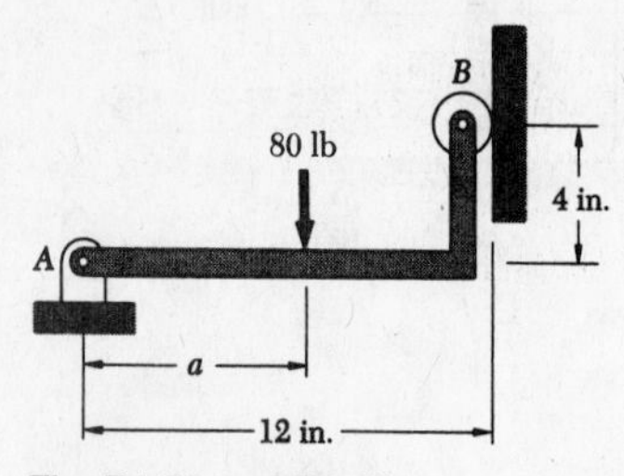

Fig. P4.39 and P4.40

4.39 Determine the reactions at A and B when $a = 7.5$ in.

4.40 Determine the value of a for which the magnitude of the reaction B is equal to 200 lb.

4.41 A 12-ft ladder, weighing 40 lb, leans against a frictionless vertical wall. The lower end of the ladder rests on rough ground, 4 ft away from the wall. Determine the reactions at both ends.

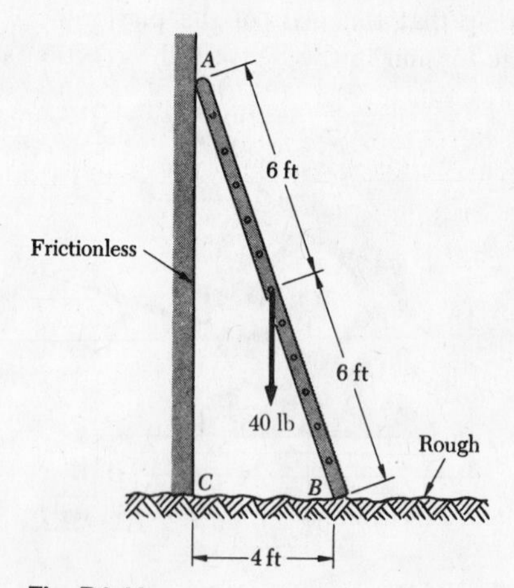

Fig. P4.41

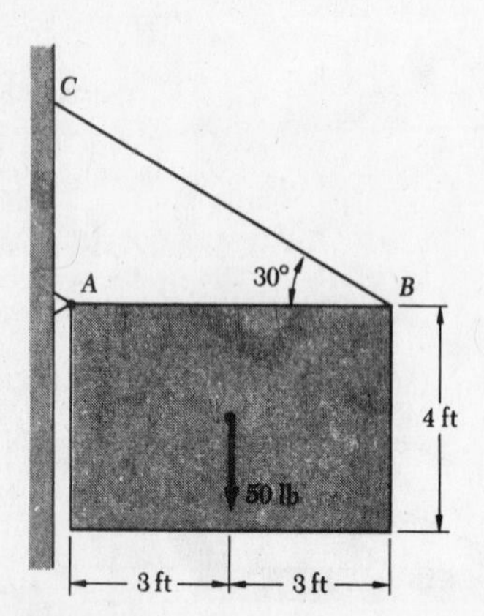

Fig. P4.42

4.42 A 50-lb sign is supported by a pin and bracket at A and by a cable BC. Determine the reaction at A and the tension in the cable.

4.43 A 100-kg roller, of diameter 500 mm, is used on a lawn. Determine the force **F** required to make it roll over a 50-mm obstruction (*a*) if the roller is pushed as shown, (*b*) if the roller is pulled as shown.

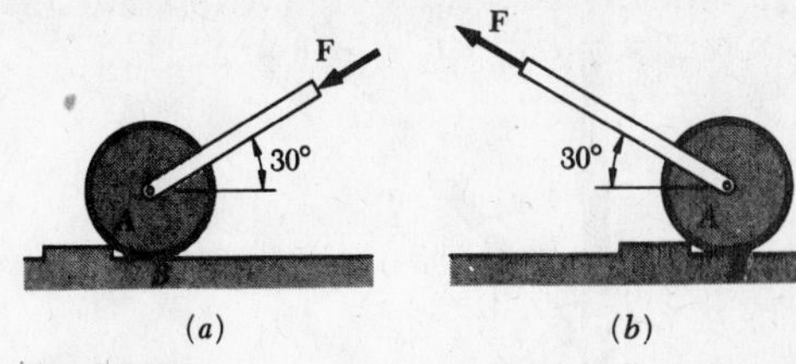

Fig. P4.43

4.44 Determine the reactions at A and E.

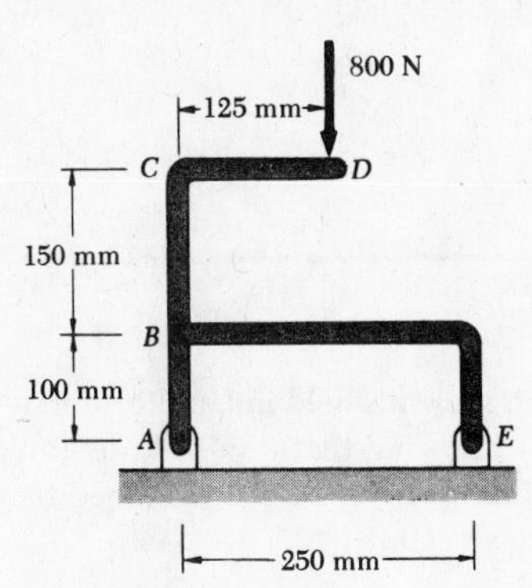

Fig. P4.44

4.45 The slender rod AB of length L and weight W is attached to a collar at A and rests on a small wheel at C. Neglecting the effect of friction and the weight of the collar, determine the angle θ corresponding to equilibrium.

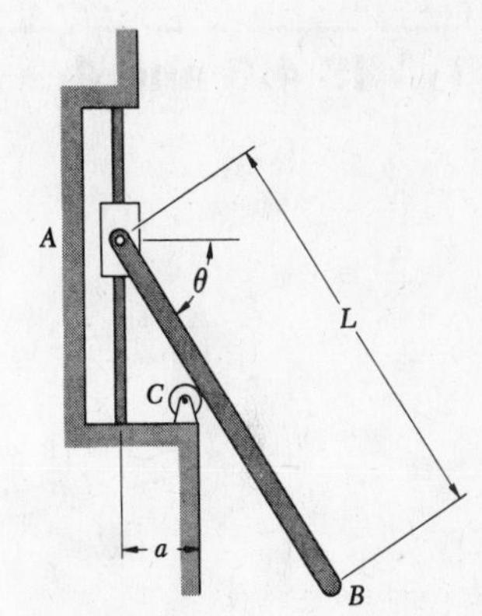

Fig. P4.45

4.46 The uniform rod AB lies in a vertical plane with its ends resting against the frictionless surfaces AC and BC. Determine the angle θ corresponding to equilibrium when (*a*) $\alpha = 30°$, (*b*) $\alpha = 40°$, (*c*) $\alpha = 60°$.

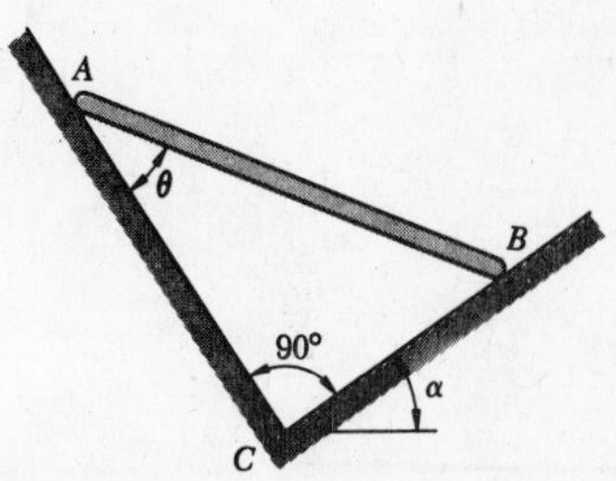

Fig. P4.46

4.47 A slender rod BC of length L and weight W is held by a cable AB and by a pin at C which may slide in a vertical slot. Knowing that $\theta = 30°$, determine (*a*) the value of β for which the rod is in equilibrium, (*b*) the corresponding tension in the cable.

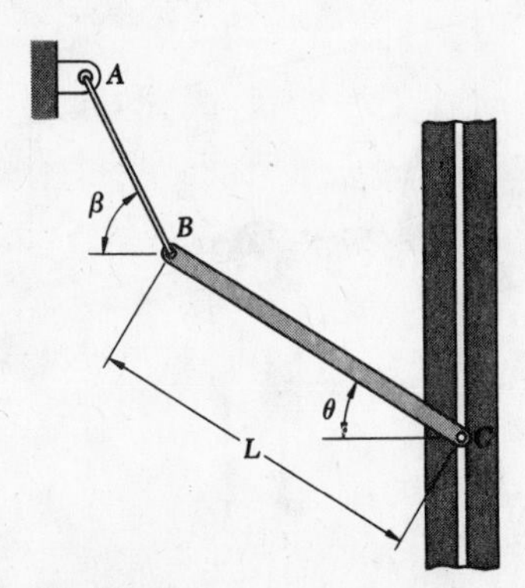

Fig. P4.47

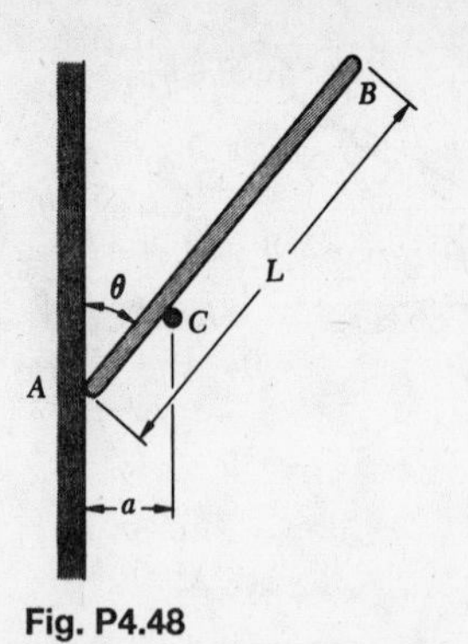

Fig. P4.48

4.48 A slender rod of length L and weight W is lodged between the wall and the peg C. Neglecting friction, determine the angle θ between the rod and the wall corresponding to equilibrium.

SECTIONS 4.8 and 4.9

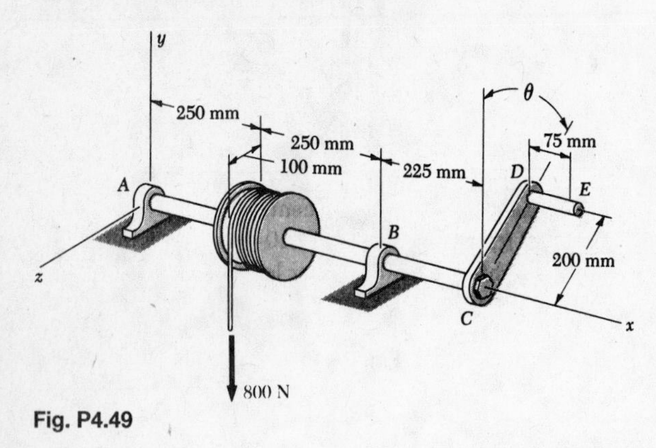

Fig. P4.49

4.49 The winch shown is held in equilibrium by a vertical force **P** applied at point E. Knowing that $\theta = 60°$, determine the magnitude of **P** and the reactions at A and B. It is assumed that the bearing at B does not exert any axial thrust.

4.50 Solve Prob. 4.49, assuming that the winch is held in equilibrium by a force **P** applied at E in a direction perpendicular to the plane CDE.

4.51 The overhead transmission shaft AE is driven at a constant speed by an electric motor connected by a flat belt to pulley B. Pulley C may be used to drive a machine tool located directly below C, while pulley D drives a parallel shaft located at the same height as AE. Knowing that $T_B + T'_B = 36$ lb, $T_C = 40$ lb, $T'_C = 16$ lb, $T_D = 0$, and $T'_D = 0$, determine (*a*) the tension in each portion of the belt driving pulley B, (*b*) the reactions at the bearings A and E caused by the tension in the belts.

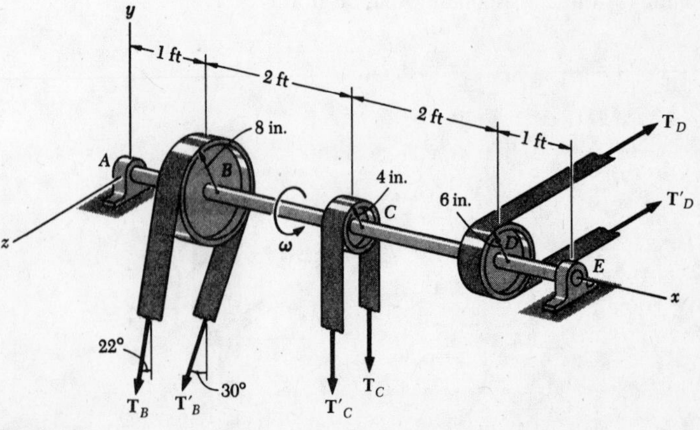

Fig. P4.51

4.52 Solve Prob. 4.51, assuming that $T_B + T'_B = 36$ lb, $T_C = 0$, $T'_C = 0$, $T_D = 36$ lb, and $T'_D = 12$ lb.

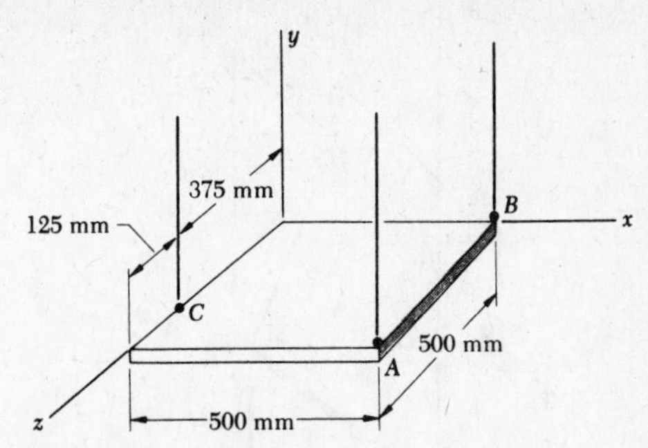

Fig. P4.53 and P4.54

4.53 The 20-kg square plate shown is supported by three wires as shown. Determine the tension in each wire.

4.54 A block of mass m is to be placed on the plate shown. Knowing that the mass of the plate is 20 kg, determine the magnitude of m and the point where the block should be placed if the tension is to be 150 N in each of the three wires.

4.55 Two steel wide-flange beams AB and BC, each of length L, are welded together at B and rest in the horizontal plane on supports at A, D, and C. Each support is mounted on a caster and can exert only an upward vertical force. Knowing that the beams weigh w lb/ft, determine the reactions at the supports if $\theta = 90°$ and $a = \frac{1}{4}L$.

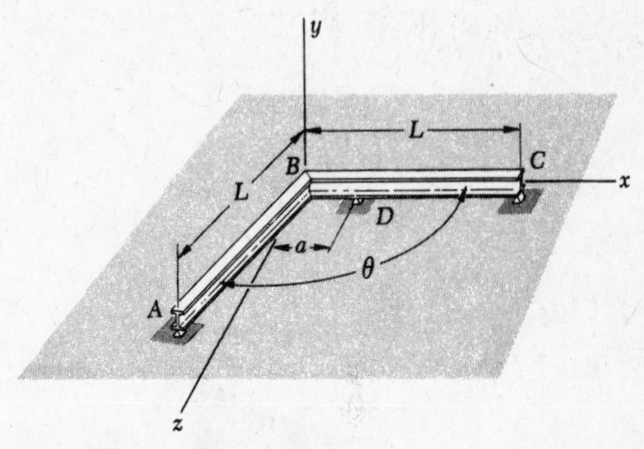

Fig. P4.55

4.56 Knowing that $\theta = 90°$, determine the largest permissible value of a if the beams of Prob. 4.55 are not to tip.

4.57 The 10-m pole is acted upon by an 8.4-kN force as shown. It is held by a ball and socket at A and by the two cables BD and BE. Neglecting the weight of the pole, determine the tension in each cable and the reaction at A.

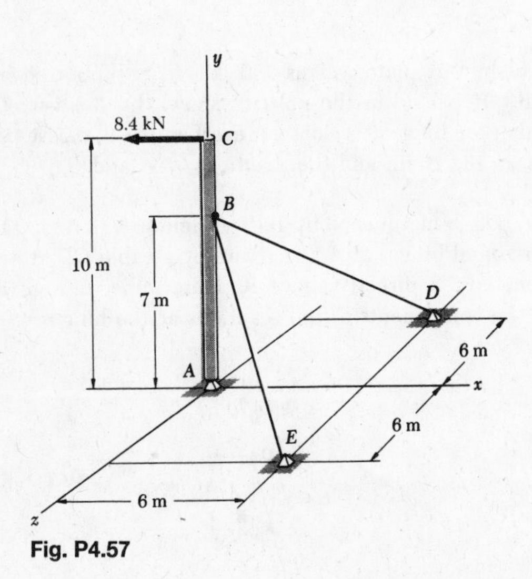

Fig. P4.57

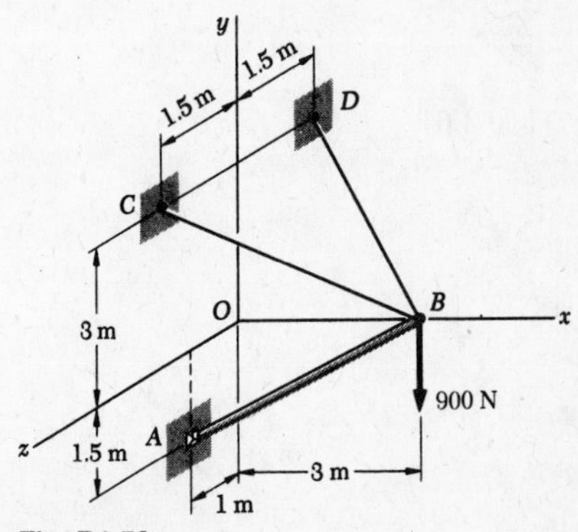

Fig. P4.58

4.58 The boom AB supports a load of 900 N as shown. The boom is held by a ball and socket at A and by the two cables BC and BD. Neglecting the weight of the boom, determine the tension in each cable and the reaction at A.

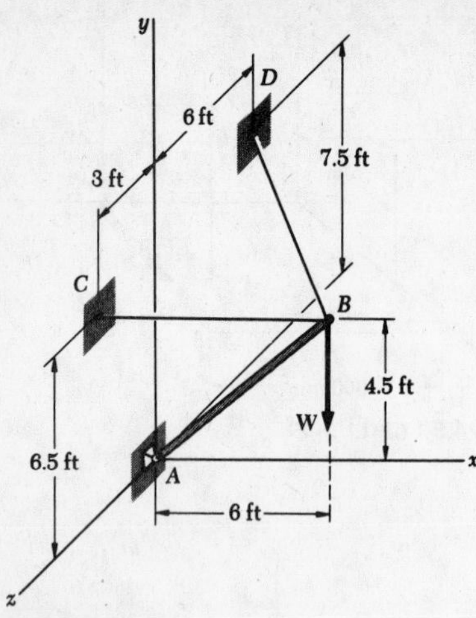

Fig. P4.59

4.59 The boom AB supports a load of weight $W = 820$ lb. The boom is held by a ball and socket at A and by the two cables BC and BD. Neglecting the weight of the boom, determine the tension in each cable and the reaction at A.

4.60 A 7-ft boom is held by a ball and socket at A and by two cables EBF and DC; cable EBF passes around a frictionless pulley at B. Determine the tension in each cable.

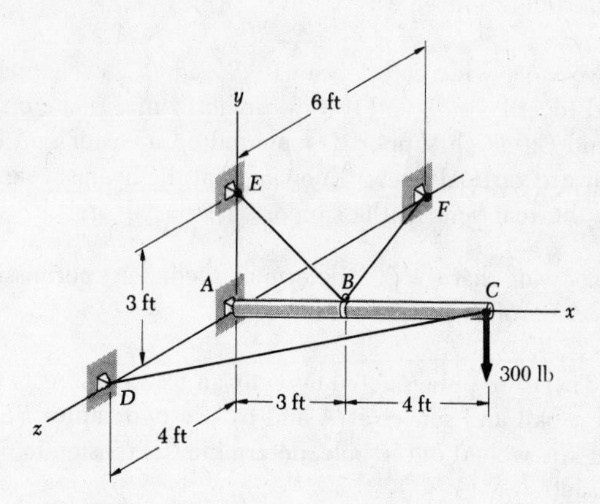

Fig. P4.60

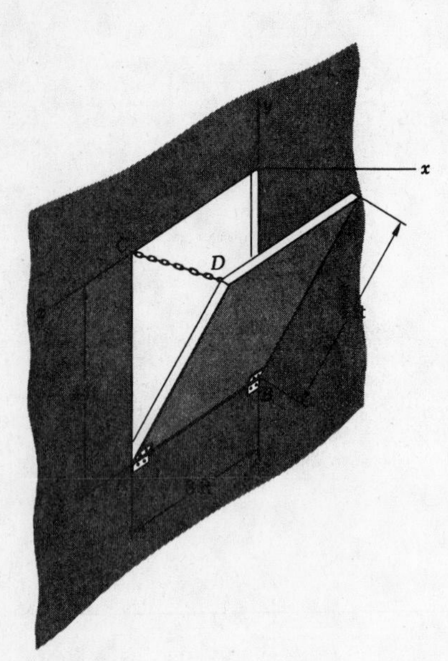

Fig. P4.61

4.61 A 3- by 4-ft plate weighs 150 lb and is supported by hinges at A and B. It is held in the position shown by the 2-ft chain CD. Assuming that the hinge at A does not exert any axial thrust, determine the tension in the chain and the reactions at A and B.

4.62 A 500-lb marquee, 8 by 10 ft, is held in a horizontal position by two horizontal hinges at A and B and by a cable CD attached to a point D located 5 ft directly above B. Determine the tension in the cable and the components of the reactions at the hinges.

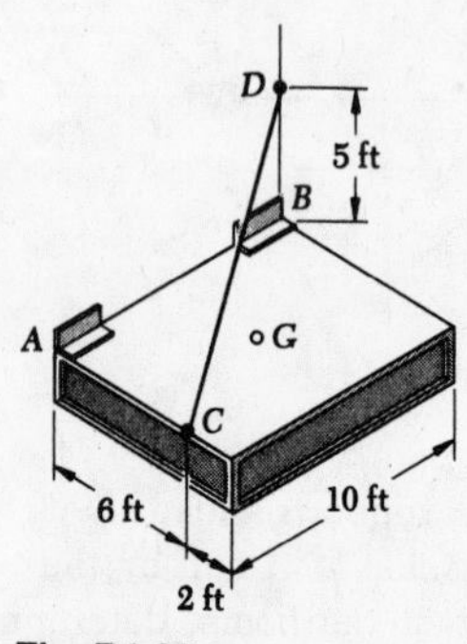

Fig. P4.62

4.63 A 10-kg storm window measuring 900 by 1500 mm is held by hinges at A and B. In the position shown, it is held away from the side of the house by a 600-mm stick CD. Assuming that the hinge at A does not exert any axial thrust, determine the magnitude of the force exerted by the stick and the components of the reactions at A and B.

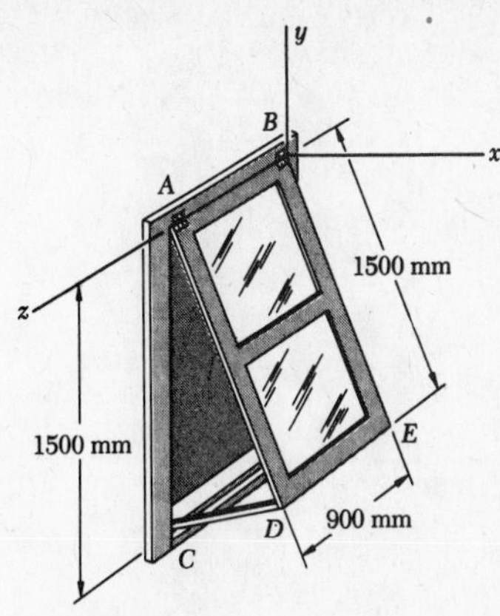

Fig. P4.63

4.64 A 20-kg door is made self-closing by hanging a 15-kg counterweight from a cable attached at C. The door is held open by a force **P** applied at the knob D, in a direction perpendicular to the door. Determine the magnitude of **P** and the components of the reactions A and B when $\theta = 90°$. It is assumed that the hinge at A does not exert any axial thrust.

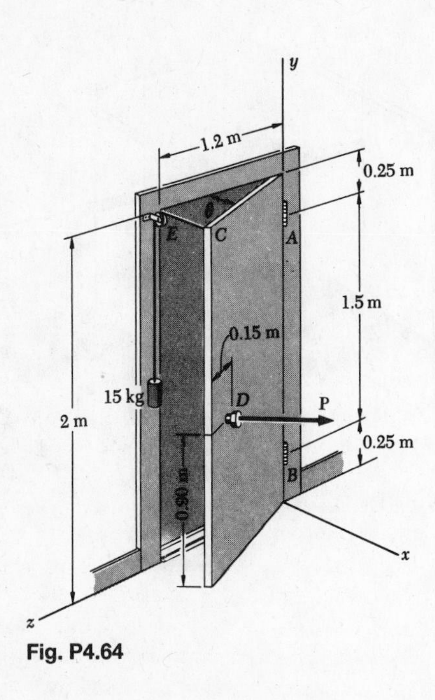

Fig. P4.64

4.65 Solve Prob 4.61, assuming that the hinge at A is removed.

4.66 Solve Prob. 4.62, assuming that the hinge at A is removed.

4.67 Solve Prob. 4.63, assuming that the hinge at A is removed.

4.68 Three rods are welded together to form the "corner" shown. The corner is supported by three smooth eyebolts. Determine the reactions at A, B, and C when $P = 1.2$ kN, $a = 300$ mm, $b = 200$ mm, and $c = 250$ mm.

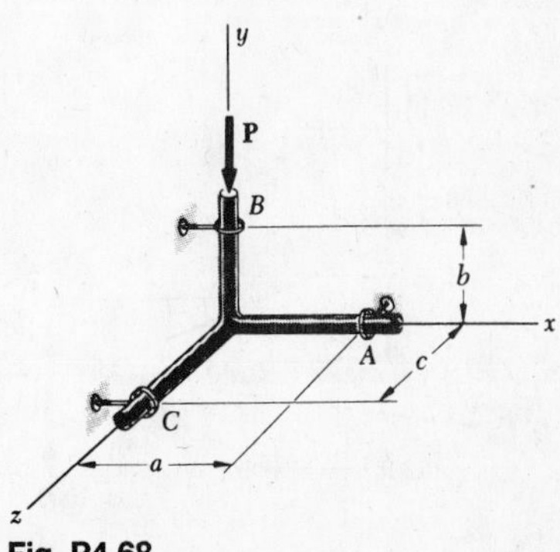

Fig. P4.68

4.69 In order to clean the clogged drainpipe AE, a plumber has disconnected both ends of the pipe and introduced a power snake through the opening at A. The cutting head of the snake is connected by a heavy cable to an electric motor which keeps it rotating at a constant speed while the plumber forces the cable into the pipe. The forces exerted by the plumber and the motor on the end of the cable may be represented by the wrench $\mathbf{F} = -(48\text{ N})\mathbf{k}$, $\mathbf{M} = -(90\text{ N}\cdot\text{m})\mathbf{k}$. Determine the additional reactions at B, C, and D caused by the cleaning operation. Assume that the reaction at each support consists of two force components perpendicular to the pipe.

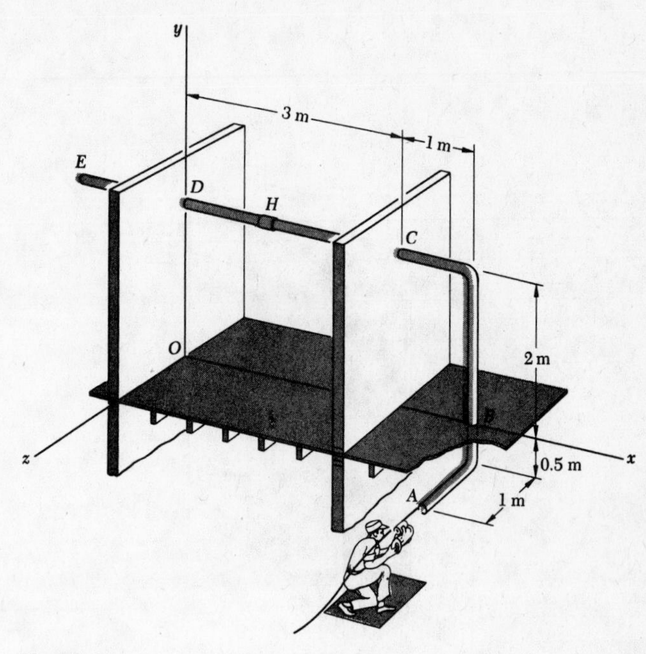

Fig. P4.69

4.70 In Prob. 4.69, the plumber has disconnected the pipe at H before introducing the snake through the opening at A. Determine the additional reactions caused by the cleaning operation at the remaining supports B and C, assuming that, in addition to the two force components perpendicular to the pipe, the reaction at C now also includes two couple vectors respectively parallel to the y and z axes.

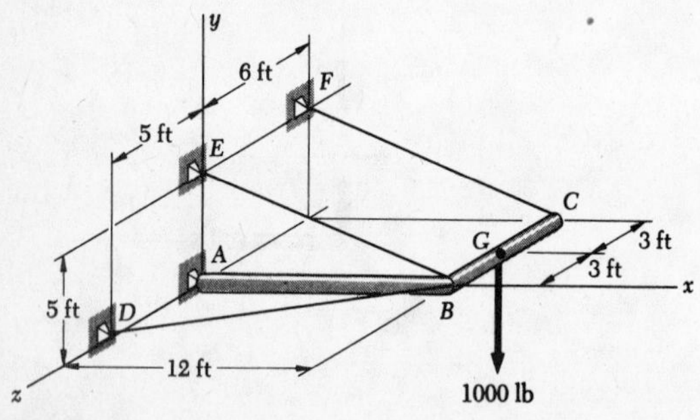

Fig. P4.71

4.71 The rigid L-shaped member ABC is supported by a ball and socket at A and by three cables. Determine the tension in each cable and the reaction at A caused by the 1000-lb load applied at G.

4.72 Solve Prob. 4.71, assuming that cable BD is removed and replaced by a cable joining points E and C.

4.73 The 8-ft rod AB and the 6-ft rod BC are hinged at B and supported by the cable DE and by ball-and-socket joints at A and C. Knowing that $h = 3$ ft, determine the tension in the cable for the loading shown.

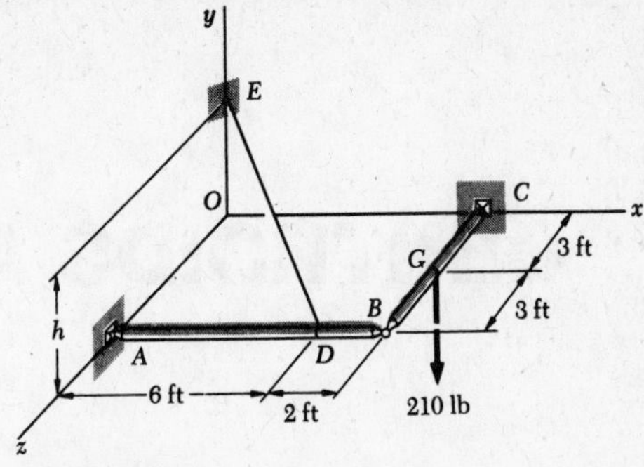

Fig. P4.73

4.74 The 23-kg plate $ABCD$ measures 325 by 450 mm; it is held by hinges along edge AD and the wire BE. Determine the tension in the wire.

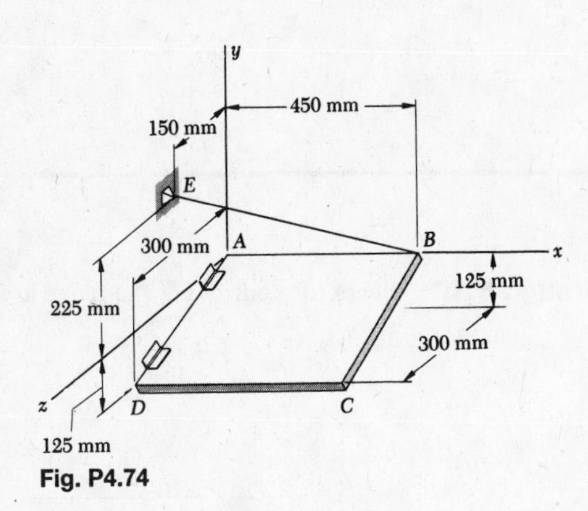

Fig. P4.74

4.75 Solve Prob. 4.74, assuming that wire BE is replaced by a wire connecting E and C.

4.76 The 60-lb load is attached to the bent bar $ABCD$. The bar is supported by the cable EC and by ball-and-socket joints at A and D. Knowing that $h = 23$ in., determine the tension in the cable.

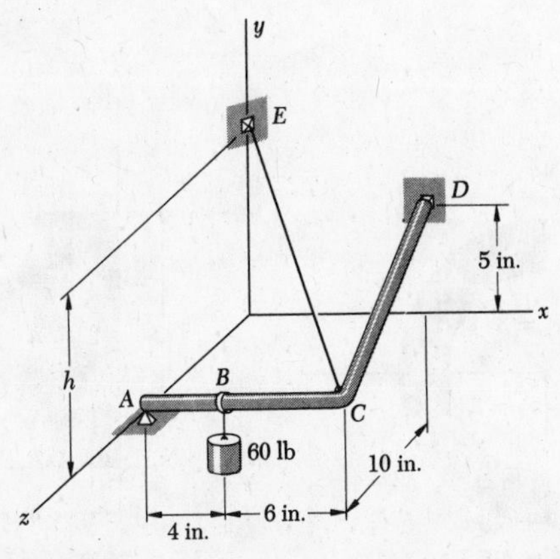

Fig. P4.76

CHAPTER 5
DISTRIBUTED FORCES: CENTROIDS AND CENTERS OF GRAVITY

SECTIONS 5.1 to 5.5

5.1 through 5.12 Locate the centroid of the plane area shown.

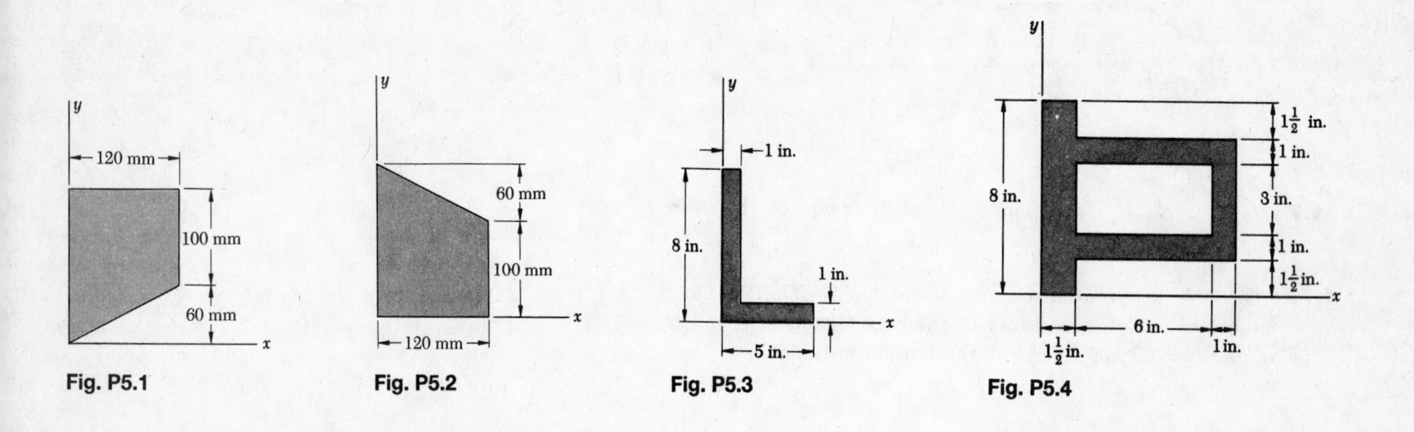

Fig. P5.1 Fig. P5.2 Fig. P5.3 Fig. P5.4

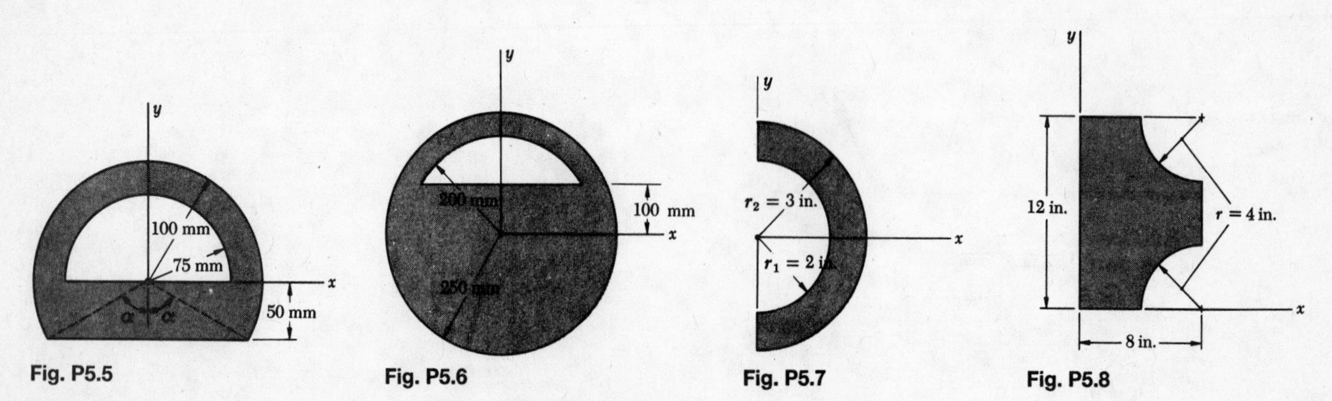

Fig. P5.5 Fig. P5.6 Fig. P5.7 Fig. P5.8

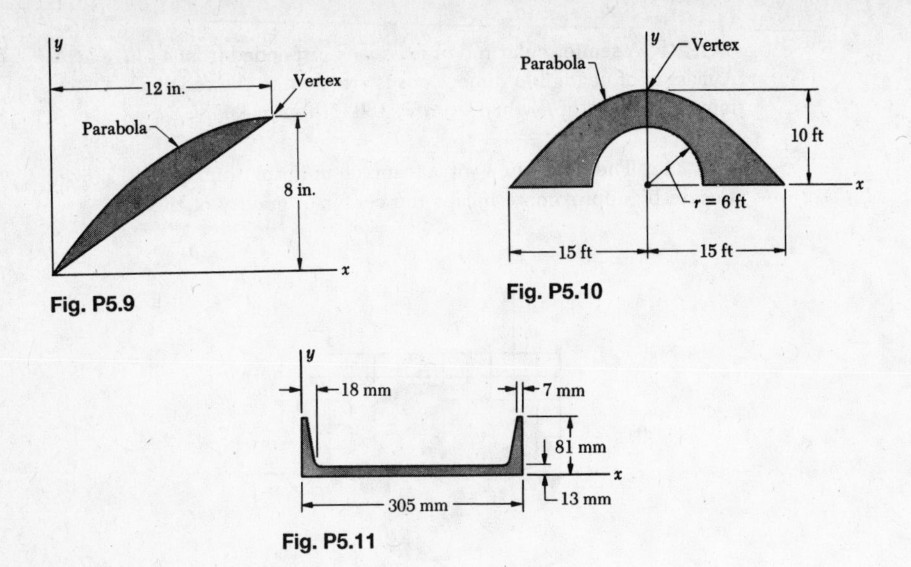

Fig. P5.9

Fig. P5.10

Fig. P5.11

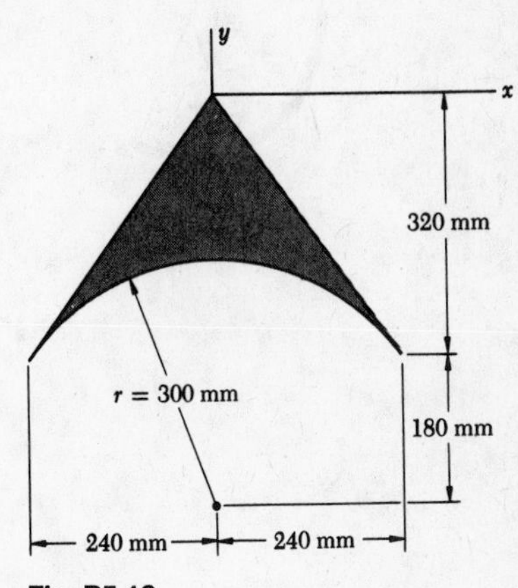

Fig. P5.12

5.13 through 5.16 A thin homogeneous wire is bent to form the *perimeter* of the figure indicated. Locate the center of gravity of the wire figure thus formed.

5.13 Fig. P5.1.
5.14 Fig. P5.2.
5.15 Fig. P5.7.
5.16 Fig. P5.8.

5.17 Knowing that the figure shown is formed of a thin homogeneous wire, determine the angle α for which the center of gravity of the figure is located at the origin O.

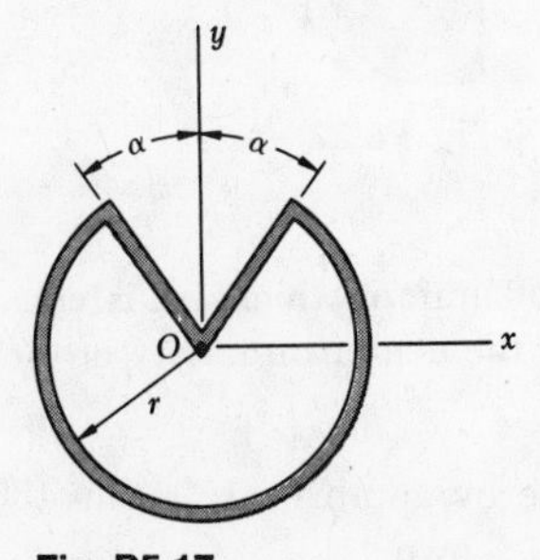

Fig. P5.17

5.18 The homogeneous wire ABC is bent as shown and is attached to a hinge at C. Determine the length L for which portion AB of the wire is horizontal.

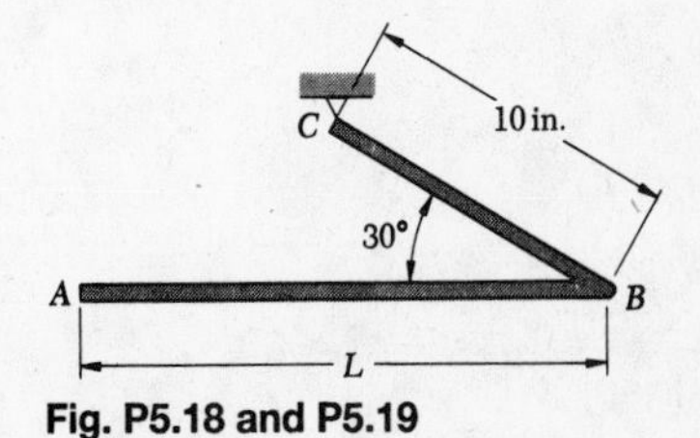

Fig. P5.18 and P5.19

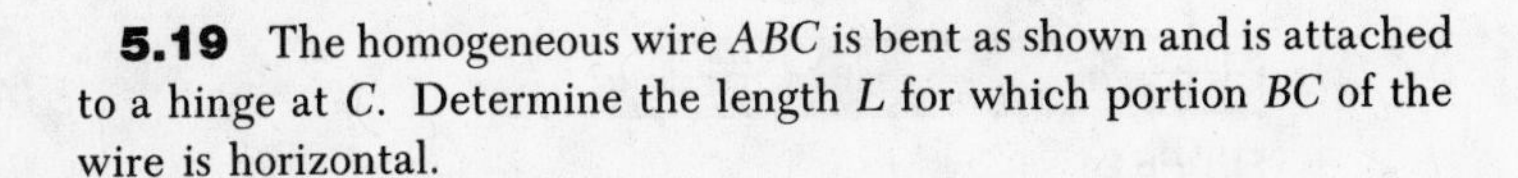

5.19 The homogeneous wire ABC is bent as shown and is attached to a hinge at C. Determine the length L for which portion BC of the wire is horizontal.

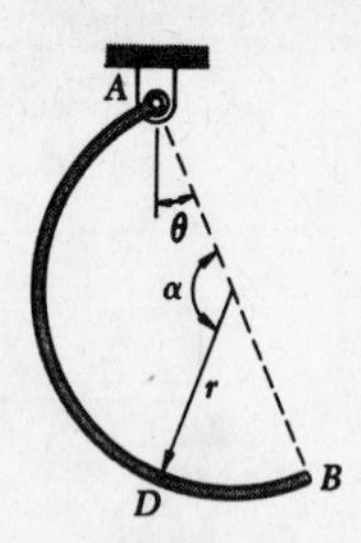

Fig. P5.20

5.20 A semicircular rod of mass m is suspended from a hinge at A. A mass m of negligible dimensions is attached to the rod at D. Determine the value of θ when (a) $\alpha = 180°$, (b) $\alpha = 90°$.

5.21 The plan view of a cam of uniform thickness is shown. Locate by approximate means the center of gravity of the cam.

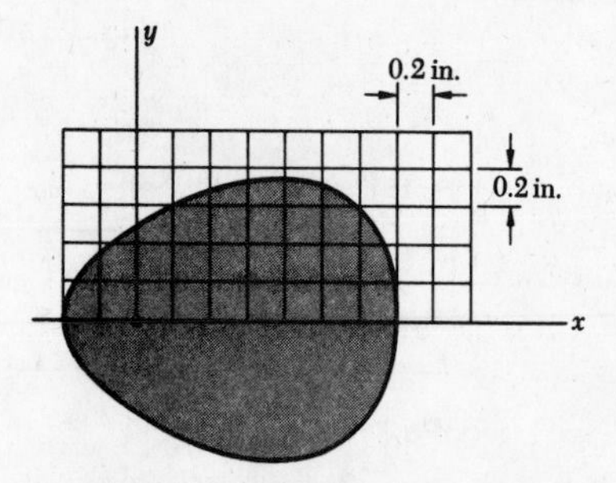

Fig. P5.21

5.22 A plate of uniform thickness is cut as shown. Locate by approximate means the center of gravity of the plate.

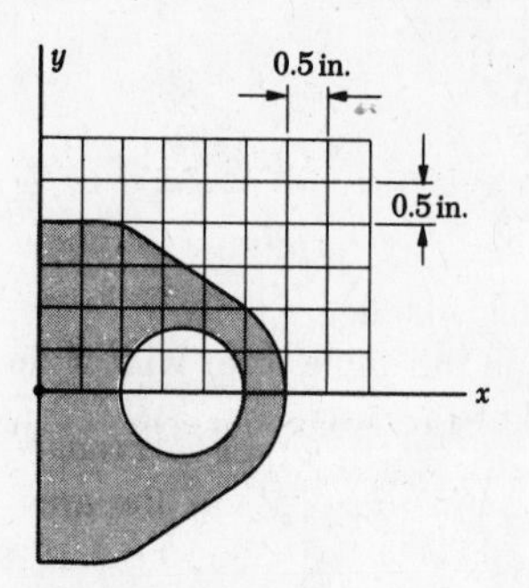

Fig. P5.22

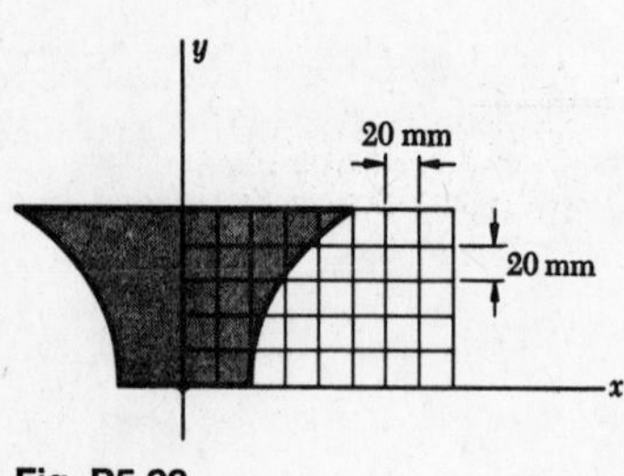

Fig. P5.23

5.23 A plate of uniform thickness is cut as shown. Locate by approximate means the center of gravity of the plate.

5.24 Determine by approximate means the x coordinate of the centroid of the area shown.

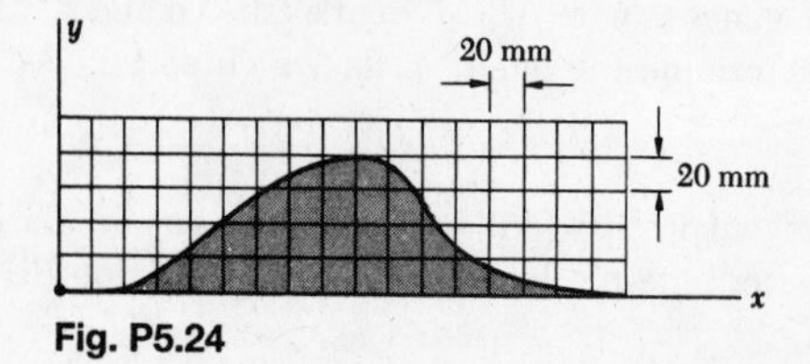

Fig. P5.24

SECTIONS 5.6 and 5.7

5.25 through 5.27 Determine by direct integration the centroid of the area shown.

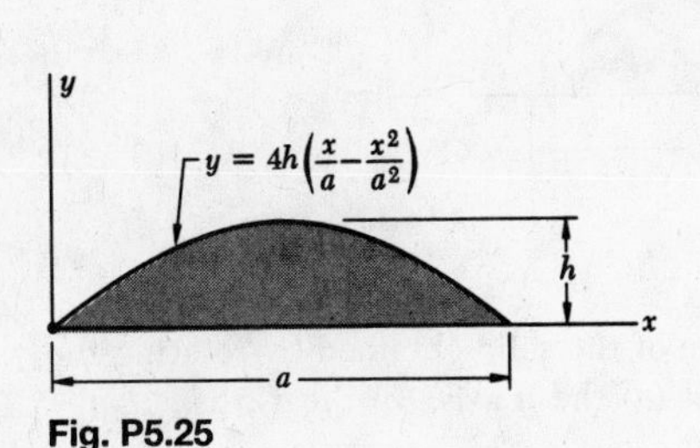

Fig. P5.25

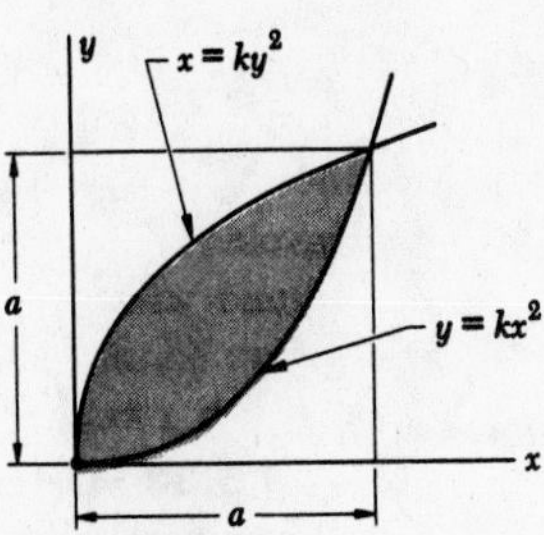

Fig. P5.26

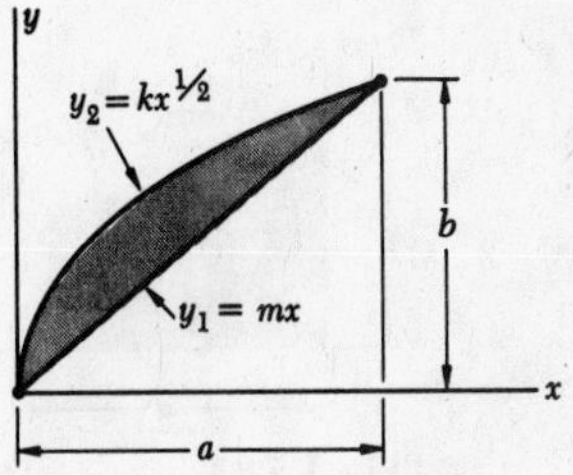

Fig. P5.27

5.28 Determine by direct integration the x coordinate of the centroid of the area shown.

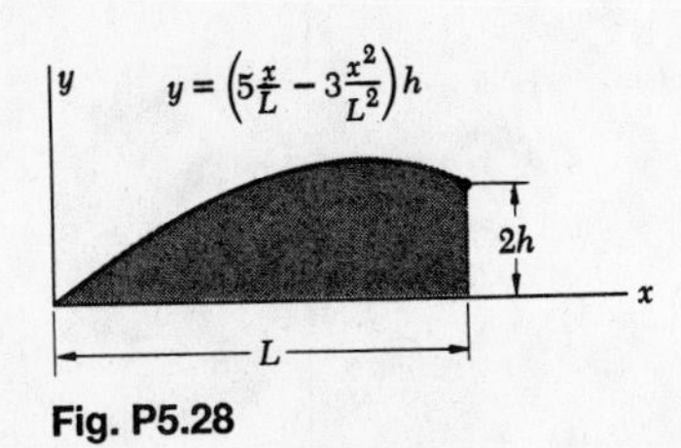

Fig. P5.28

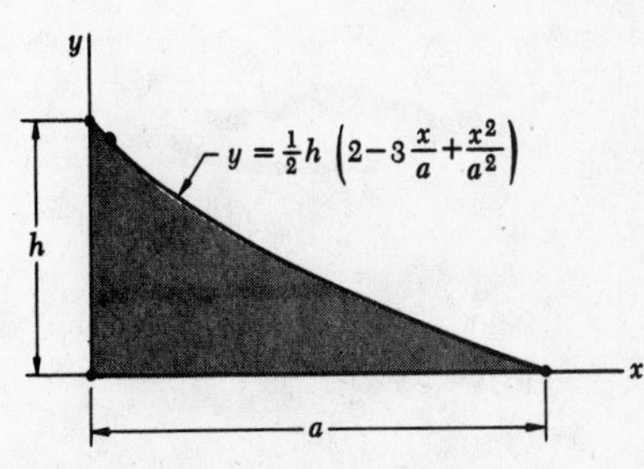

Fig. P5.29 and P5.30

5.29 Determine by direct integration the x coordinate of the centroid of the area shown.

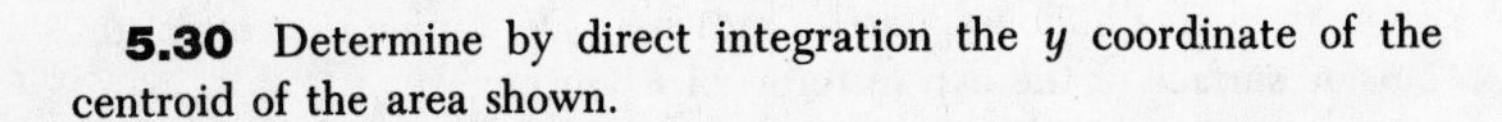

5.30 Determine by direct integration the y coordinate of the centroid of the area shown.

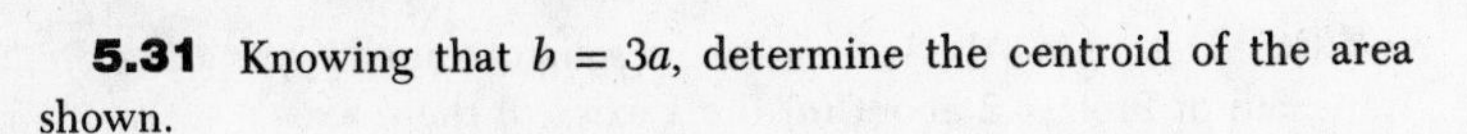

5.31 Knowing that $b = 3a$, determine the centroid of the area shown.

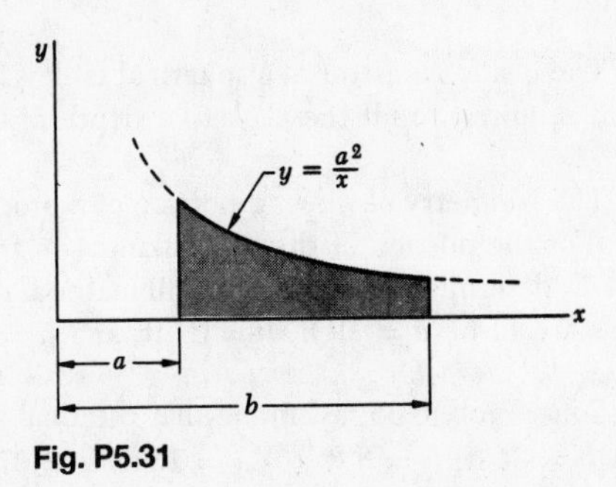

Fig. P5.31

***5.32** Determine by direct integration the centroid of the area shown.

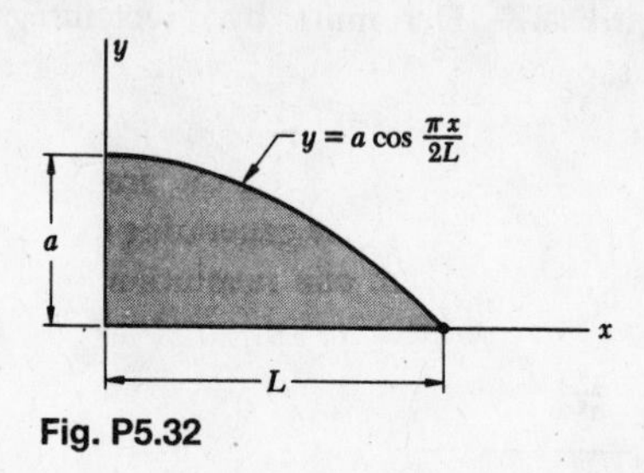

Fig. P5.32

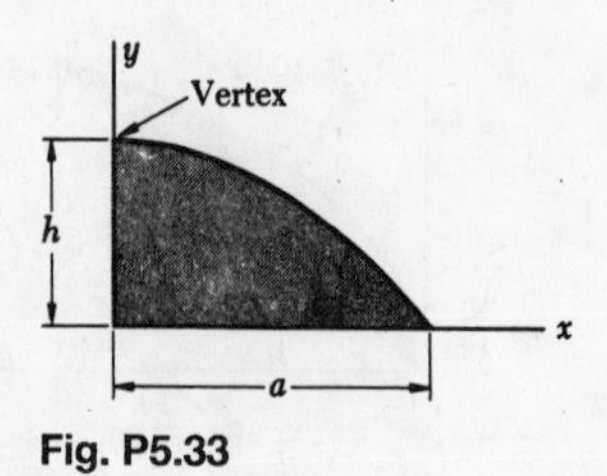

Fig. P5.33

5.33 Determine the volume of the solid obtained by rotating the semiparabolic area shown about (*a*) the y axis, (*b*) the x axis.

5.34 Determine the surface area and the volume of the half-torus shown.

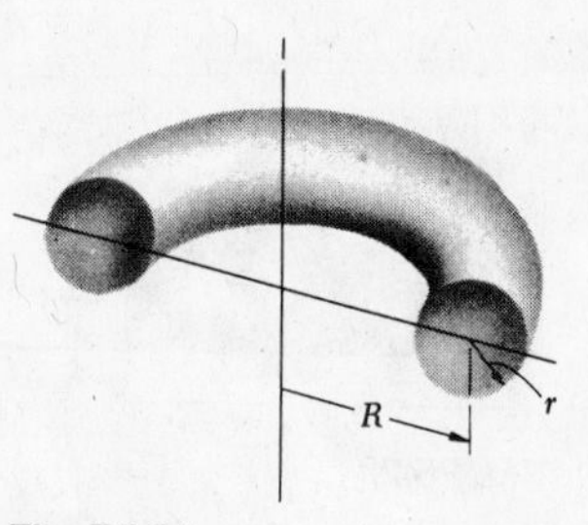

Fig. P5.34

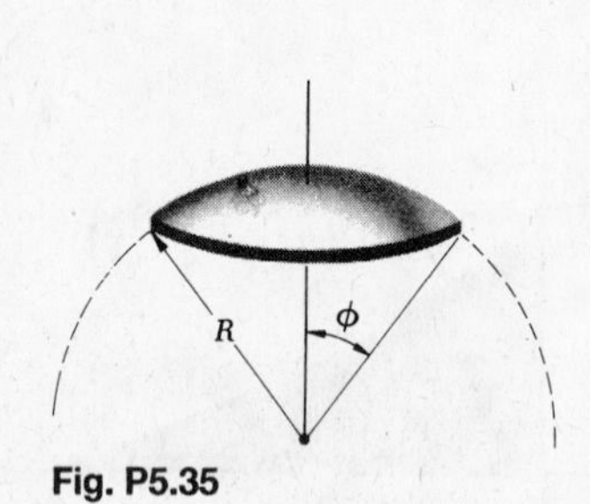

Fig. P5.35

5.35 The spherical cap shown is formed by passing a horizontal plane through a hollow sphere of radius R. Determine the area of the outside surface of the cap in terms of R and ϕ.

5.36 Determine the volume of the solid obtained by rotating the trapezoid of Prob. 5.2 about (*a*) the x axis, (*b*) the y axis.

5.37 The inside diameter of a spherical tank is 2 m. What volume of liquid is required to fill the tank to a depth of 0.5 m?

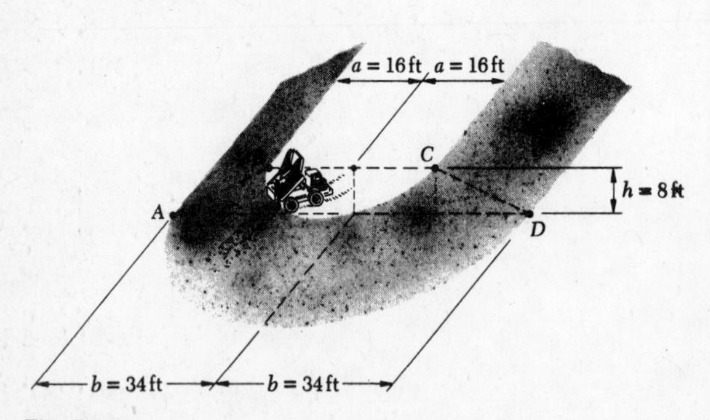

Fig. P5.38

5.38 The geometry of the end portion of a proposed embankment is shown. For the portion of the embankment in front of the vertical plane $ABCD$, determine the additional fill material required if the final dimensions are to be $a = 19$ ft, $b = 37$ ft, and $h = 8$ ft.

5.39 Solve Prob. 5.38, assuming that the final dimensions are (*a*) $a = 19$ ft, $b = 34$ ft, $h = 8$ ft, (*b*) $a = 16$ ft, $b = 37$ ft, $h = 8$ ft.

5.40 Determine the volume of the steel collar obtained by rotating the shaded area shown about the vertical axis AA'.

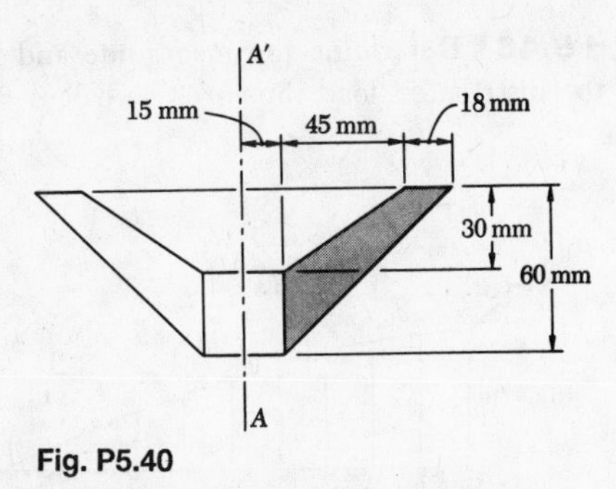

Fig. P5.40

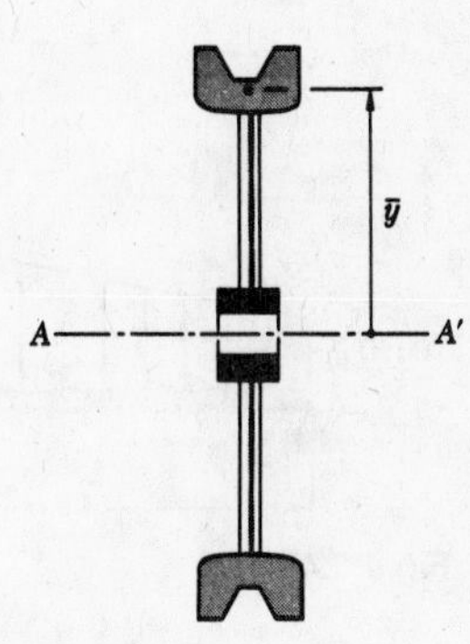

Fig. P5.41

5.41 The rim of a steel V-belt pulley has a mass of 3.9 kg. Knowing that the area of the cross section of the rim is 522 mm^2, determine the distance $\bar{y}$ from the axle AA' to the centroid of the cross-sectional area of the rim. (Density of steel = 7850 kg/m^3.)

5.42 Determine the volume of material removed as the rod is sharpened to the shape indicated by the dashed lines.

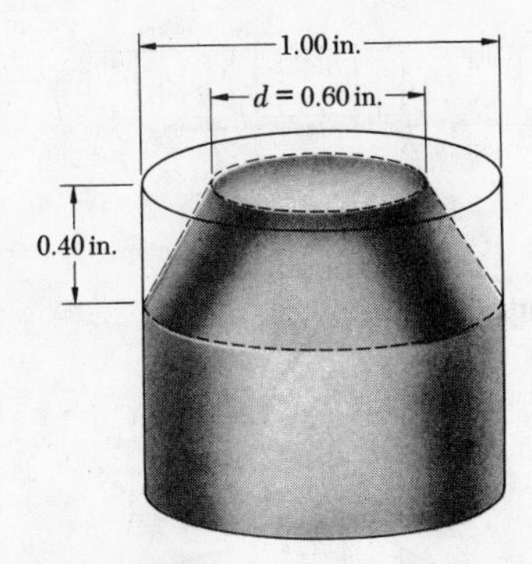

Fig. P5.42

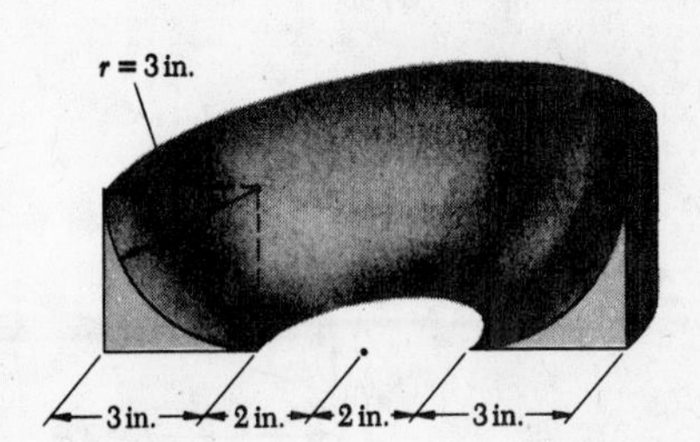

Fig. P5.43

5.43 Determine (*a*) the volume of the body shown, (*b*) the area of its inside curved surface.

5.44 Determine the volume and total surface area of the portion of ring shown. The cross section of the ring is a semicircle.

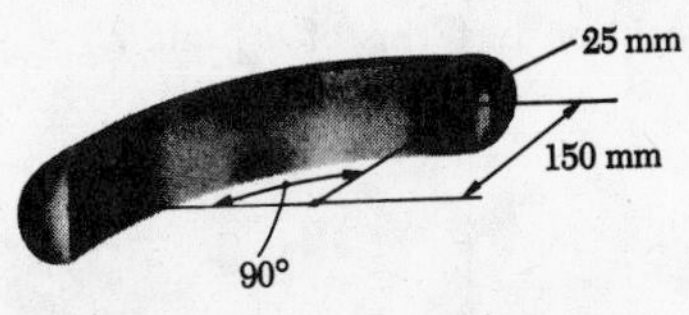

Fig. P5.44

SECTIONS 5.8 and 5.9

5.45 and 5.46 Determine the magnitude and location of the resultant of the distributed load shown. Also calculate the reactions at A and B.

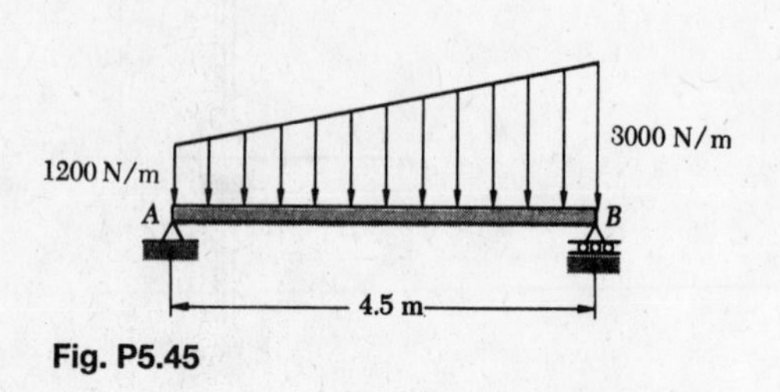

Fig. P5.45

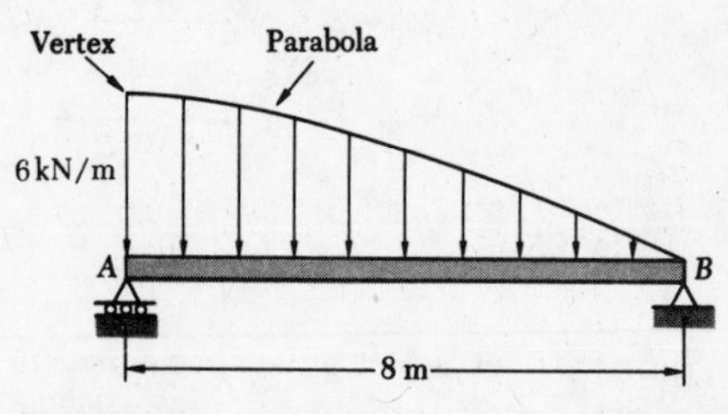

Fig. P5.46

5.47 through 5.50 Determine the reactions at the beam supports for the given loading condition.

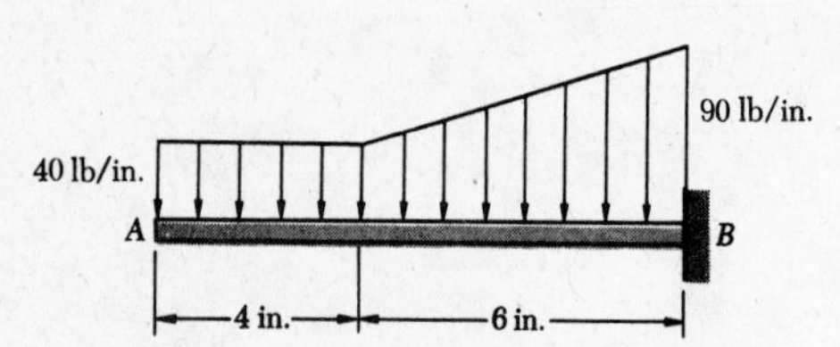

Fig. P5.47

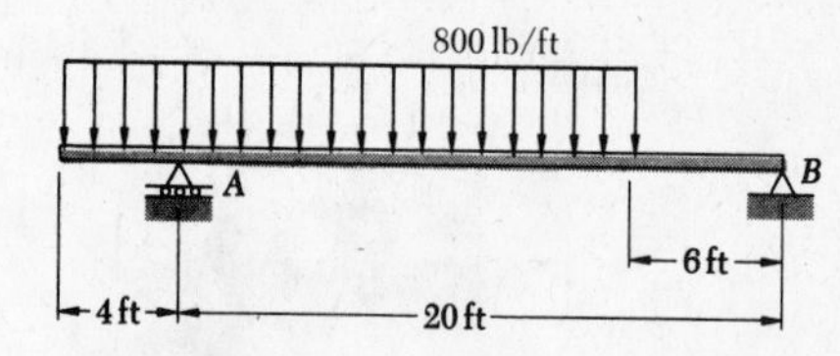

Fig. P5.48

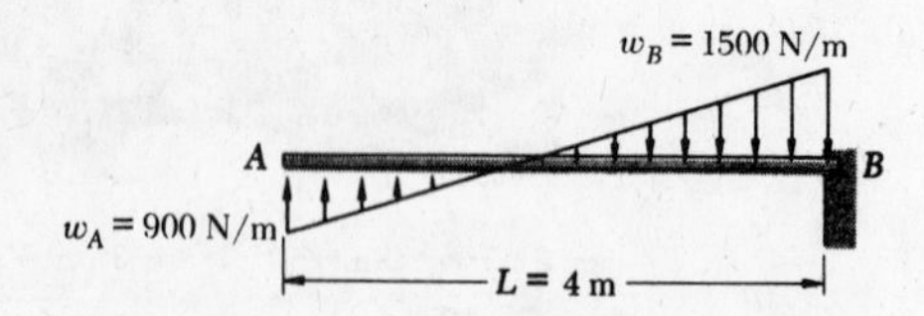

Fig. P5.49

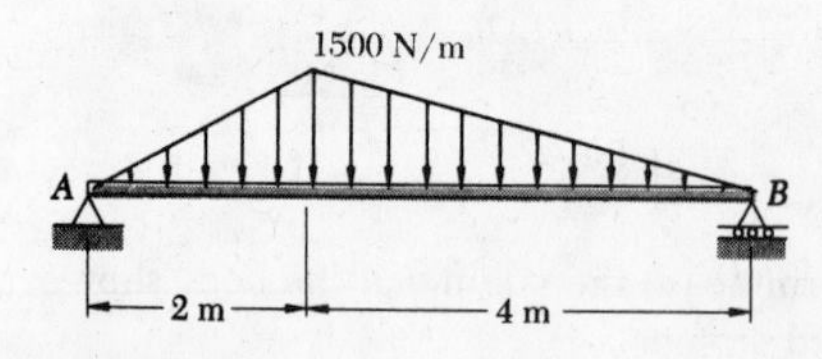

Fig. P5.50

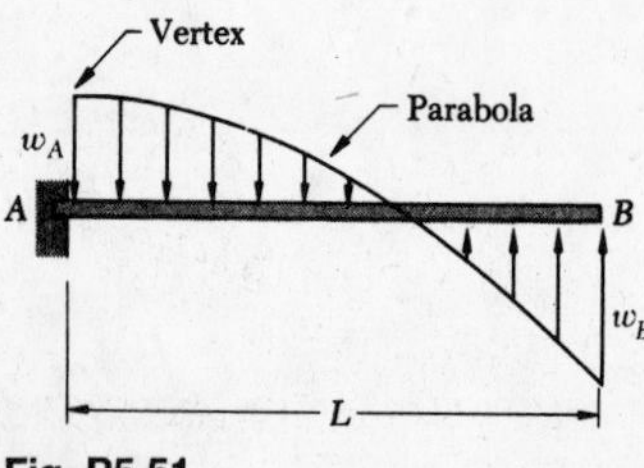

Fig. P5.51

5.51 Determine the ratio of w_A to w_B for which the reaction at A is equal to (*a*) a couple and no force, (*b*) a force and no couple. In each case express the reaction in terms of w_A and L.

5.52 A beam supports a uniformly distributed load w_1 and rests on soil which exerts a uniformly varying upward load as shown. (*a*) Determine w_2 and w_3, corresponding to equilibrium. (*b*) Knowing that at any point the soil can exert only an upward loading on the beam, state for what range of values of a/L the results obtained are valid.

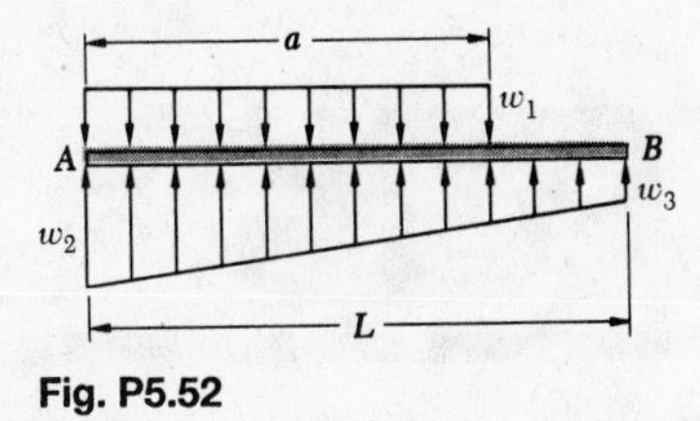

Fig. P5.52

5.53 The cross section of a concrete dam is as shown. Consider a section of the dam 1 ft thick, and determine (*a*) the resultant of the reaction forces exerted by the ground on the base of the dam *AB*, (*b*) the resultant of the pressure forces exerted by the water on the face *BC* of the dam.

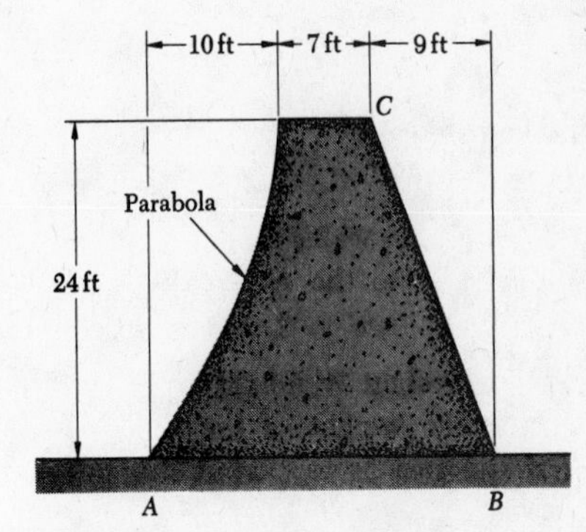

Fig. P5.53

5.54 An automatic valve consists of a square plate, 225 by 225 mm, which is pivoted about a horizontal axis through *A* located at a distance h = 100 mm above the lower edge. Determine the depth of water *d* for which the valve will open.

5.55 If the valve shown is to open when the depth of water is d = 300 mm, determine the distance *h* from the bottom of the valve to the pivot *A*.

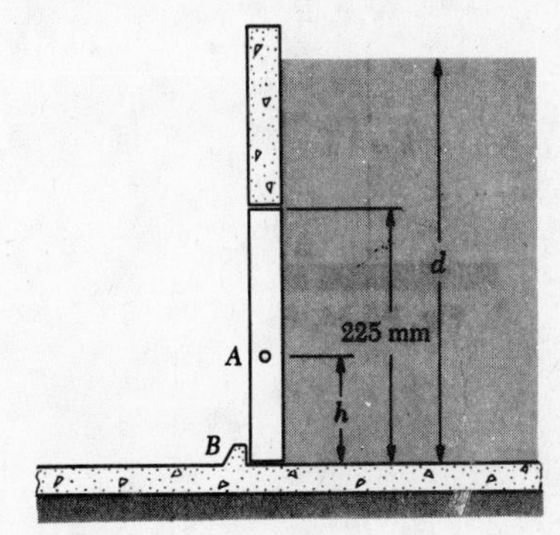

Fig. P5.54 and P5.55

5.56 A 3- by 3-ft gate is placed in a wall below water level as shown. (*a*) Determine the magnitude and location of the resultant of the forces exerted by the water on the gate. (*b*) If the gate is hinged at *A*, determine the force exerted by the sill on the gate at *B*.

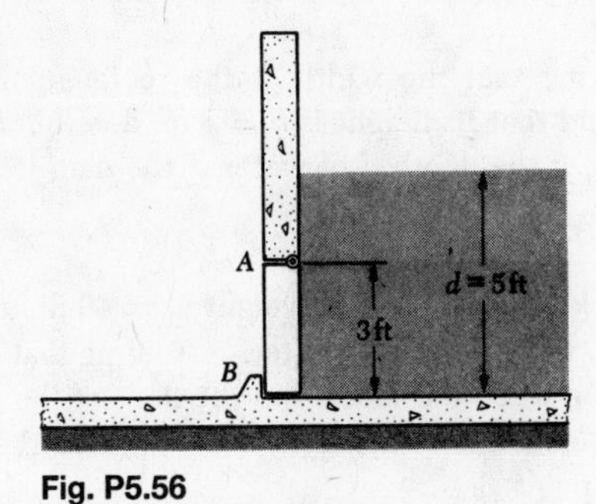

Fig. P5.56

5.57 The bent plate *ABCD* is 2 m wide and is hinged at *A*. Determine the reactions at *A* and *D* for the water level shown.

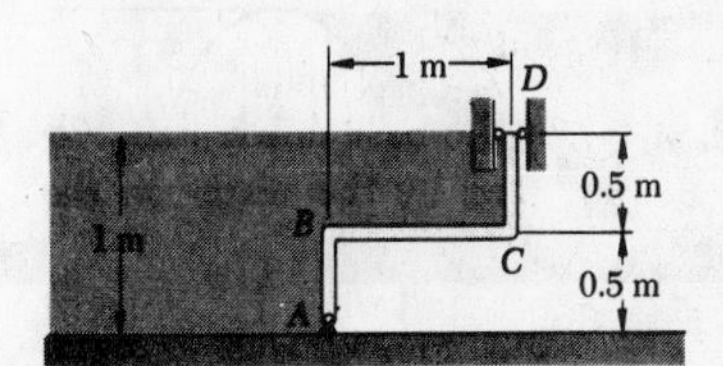

Fig. P5.57

5.58 A uniform rectangular gate of weight W, height r, and length b is hinged at A. Denoting the specific weight of the fluid by γ, determine the required angle θ if the gate is to permit flow when $d = r$.

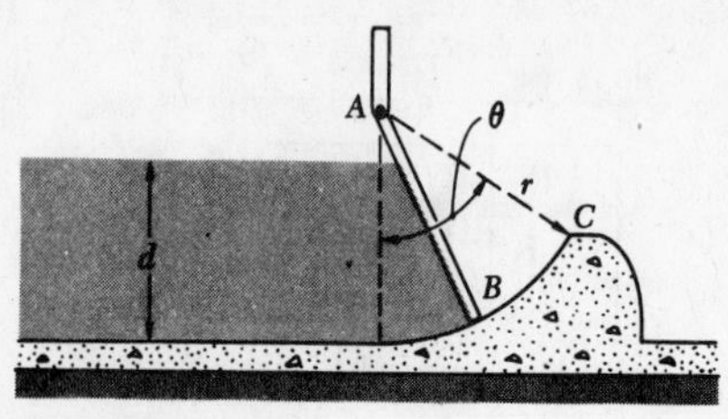

Fig. P5.58

5.59 The end of a freshwater channel consists of a plate ABC which is hinged at B and is 3 ft wide. Knowing that $b = 2$ ft and $h = 1.5$ ft, determine the reactions at A and B.

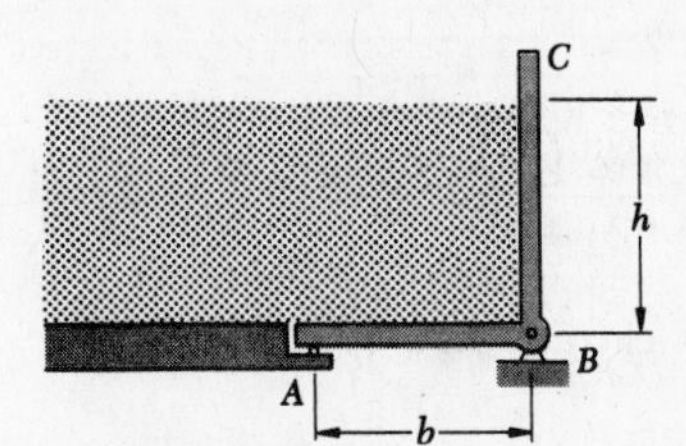

Fig. P5.59 and P5.60

5.60 The end of a freshwater channel consists of a plate ABC which is hinged at B and is 3 ft wide. Determine the ratio h/b for which the reaction at A is zero.

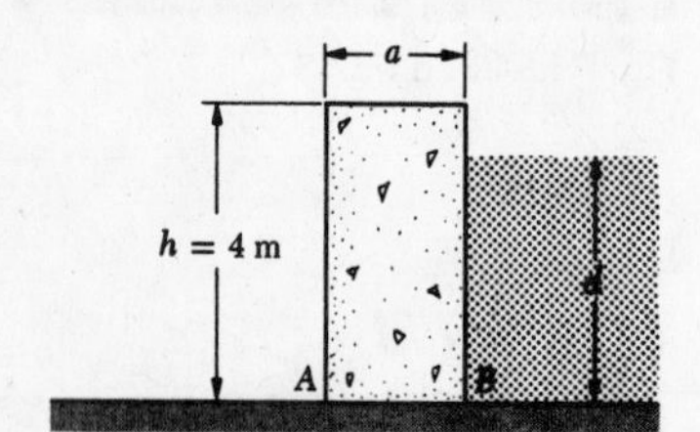

Fig. P5.61 and P5.62

5.61 Determine the minimum allowable value of the width a of the rectangular concrete dam if the dam is not to overturn about point A when $d = h = 4$ m.

5.62 Knowing that the width of the rectangular concrete dam is $a = 1.25$ m and that its height is $h = 4$ m, determine the maximum allowable value of the depth d of water if the dam is not to overturn about A.

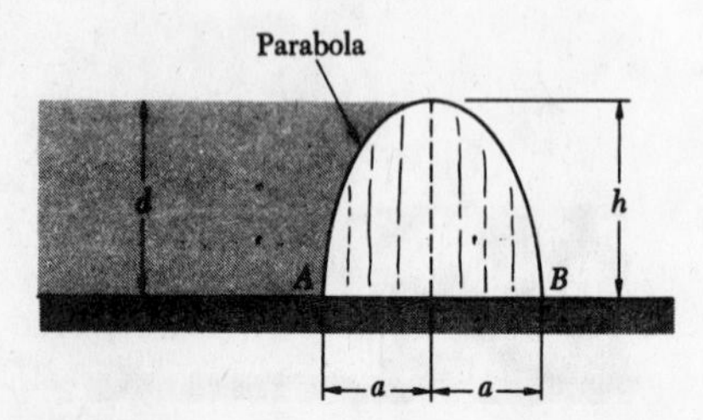

Fig. P5.63

5.63 A block of wood (specific weight $\gamma_1 = 40\ \text{lb/ft}^3$) is placed in a small channel to stop the flow of water. Assuming that $d = h$ and that no water leaks between the block and the floor of the channel, determine the maximum value of the ratio h/a for which the block will not overturn about point B.

5.64 Solve Prob. 5.63, assuming that leakage occurs under the block, causing an upward pressure on the base which varies linearly from zero at B to the full hydrostatic pressure at A.

5.65 A hemisphere and a cone are attached as shown. Determine the ratio h/a for which the centroid of the composite body is located in the plane between the hemisphere and the cone.

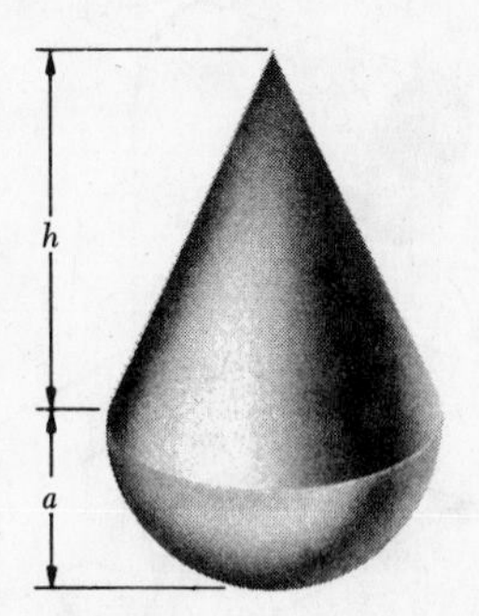

Fig. P5.65

5.66 A hemisphere and a cylinder are placed together as shown. Determine the ratio h/r for which the centroid of the composite body is located in the plane between the hemisphere and the cylinder.

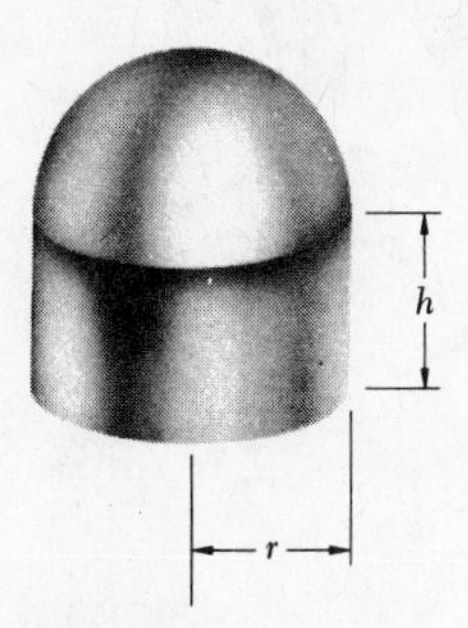

Fig. P5.66

5.67 A cone and a cylinder of the same radius a and height h are attached as shown. Determine the location of the centroid of the composite body.

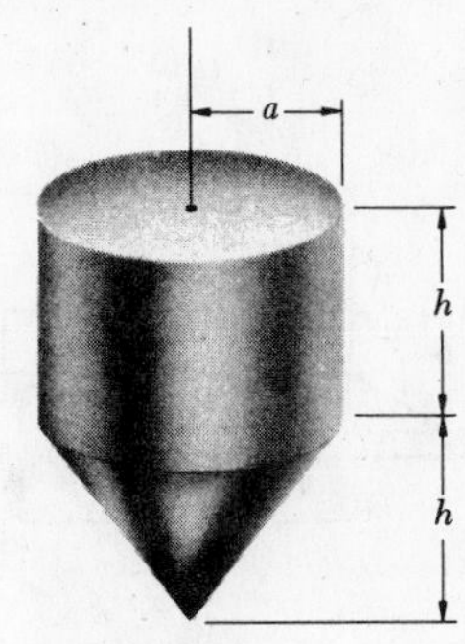

Fig. P5.67

5.68 Locate the centroid of the frustum of a right circular cone when $r_1 = 100$ mm, $r_2 = 125$ mm, and $h = 150$ mm.

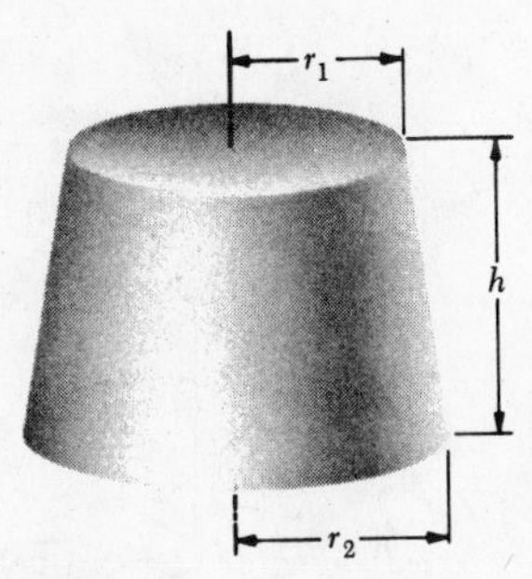

Fig. P5.68

5.69 Locate the center of gravity of the machine element shown.

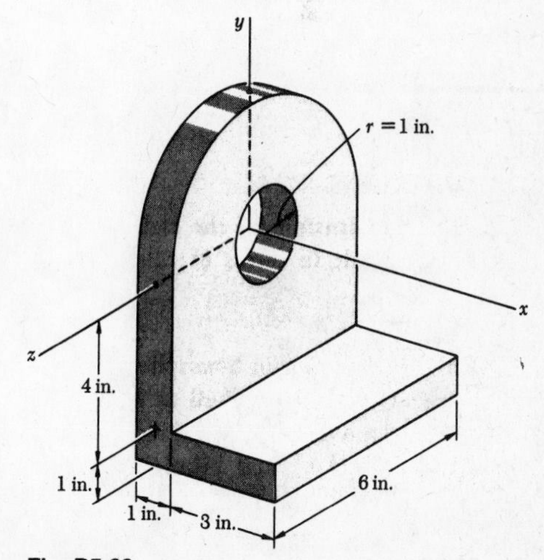

Fig. P5.69

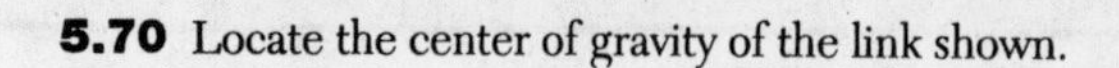

5.70 Locate the center of gravity of the link shown.

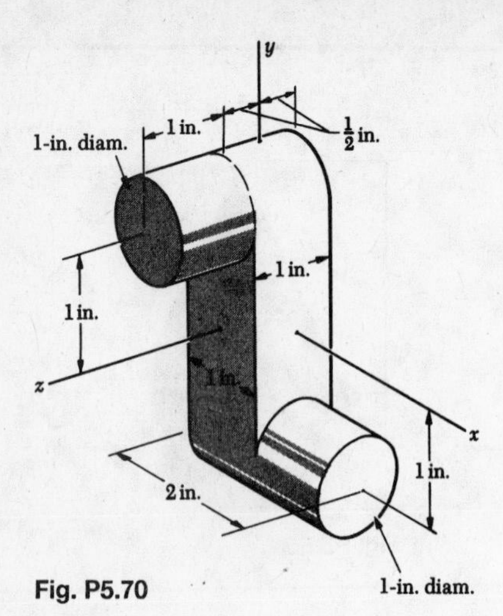

Fig. P5.70

5.71 Locate the center of gravity of the machine element shown.

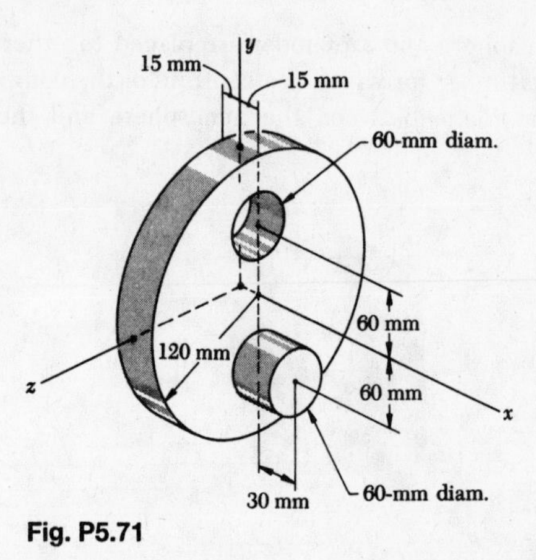

Fig. P5.71

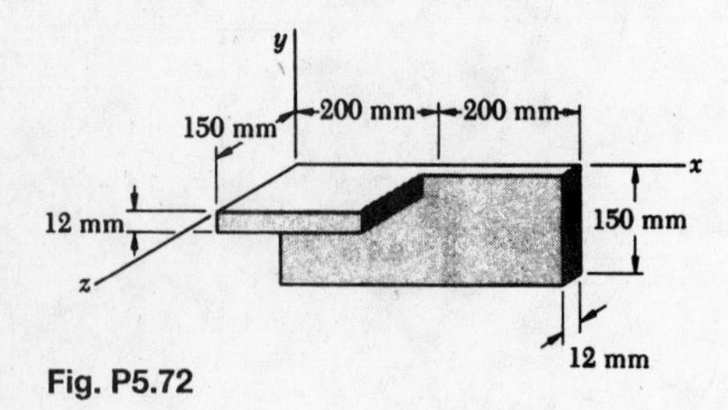

Fig. P5.72

5.72 A portion of one leg of a 150- by 150- by 12 mm angle is cut off. Determine the coordinates of the center of gravity of the remaining portion of the angle.

5.73 through 5.75 Locate the center of gravity of the sheet-metal form shown.

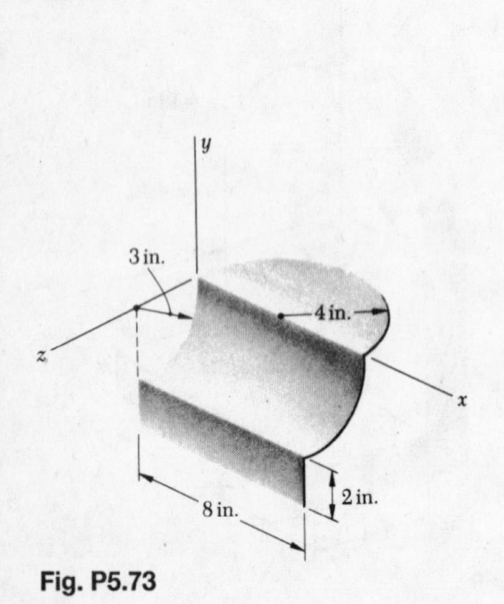

Fig. P5.73

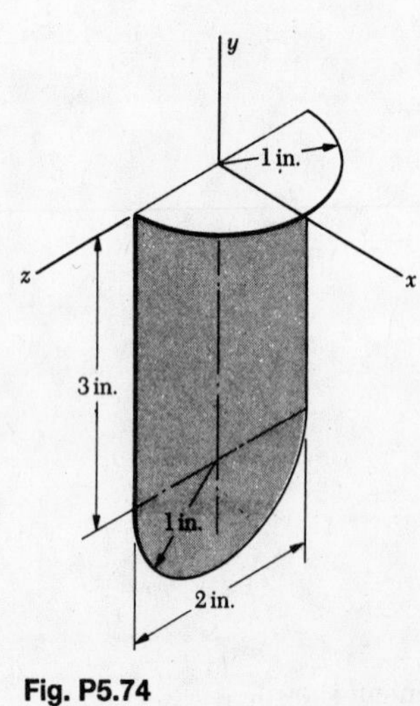

Fig. P5.74

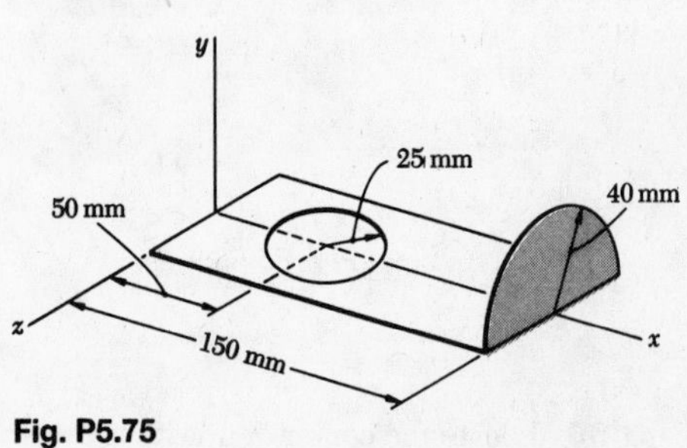

Fig. P5.75

5.76 Locate the center of gravity of the sheet-metal form shown when $a = 0.2$ m.

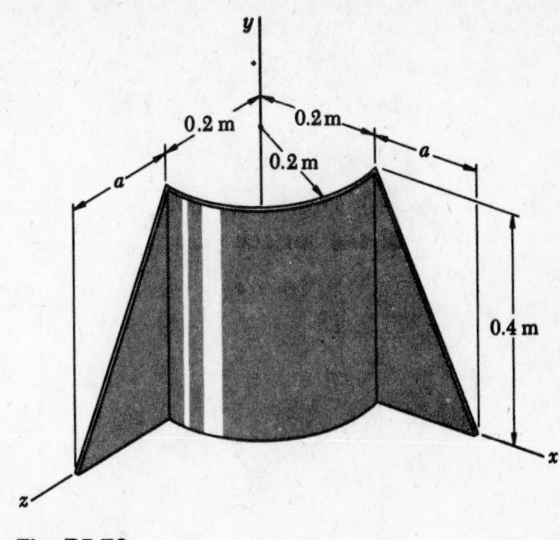

Fig. P5.76

5.77 Concrete is poured into a steel pipe of length 2 ft, inside diameter 1 ft, and weight 100 lb. To what depth should concrete be poured if the combined center of gravity of the pipe and concrete is to be as low as possible? (Specific weight of concrete = 150 lb/ft^3.)

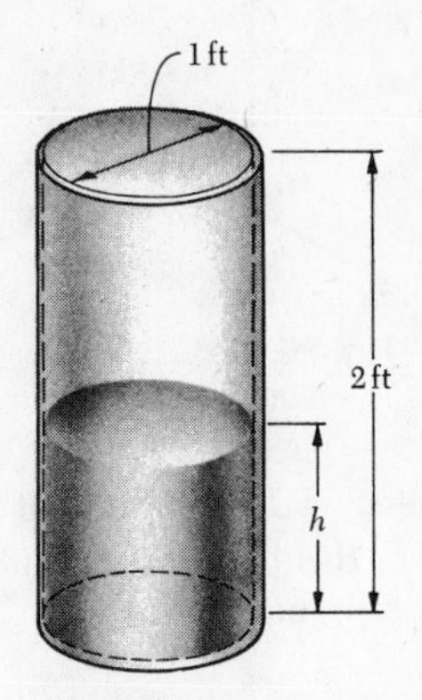

Fig. P5.77

5.78 A brass collar, of length 2 in., is mounted on an aluminum bar of length 6 in. Locate the center of gravity of the composite body. (Specific weights: brass = 530 lb/ft^3; aluminum = 170 lb/ft^3.)

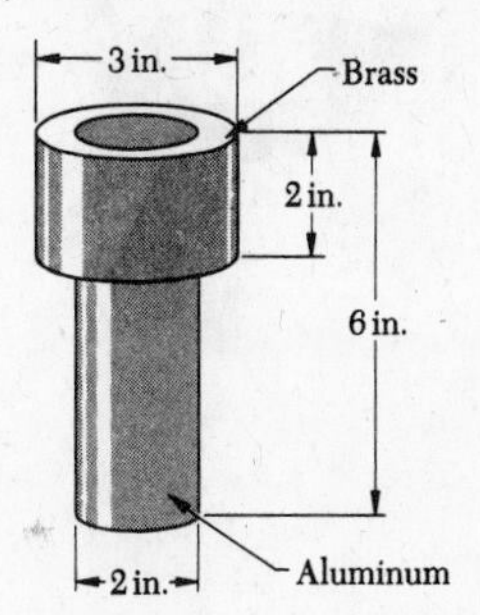

Fig. P5.78

5.79 The brass sleeve is to be mounted on the pin of a machine part made of aluminum. Locate the center of gravity of the assembly. (Specific weights: brass = 530 lb/ft^3; aluminum = 170 lb/ft^3.)

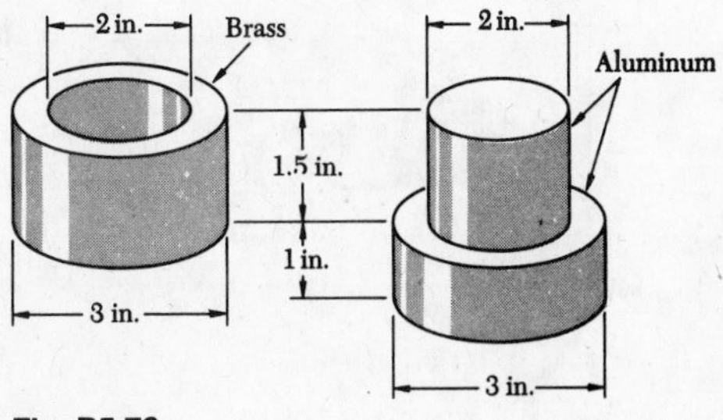

Fig. P5.79

5.80 A regular pyramid 300 mm high, with a square base of side 250 mm, is made of wood. Its four triangular faces are covered with steel sheets 1 mm thick. Locate the center of gravity of the composite body. (Densities: steel = 7850 kg/m^3; wood = 550 kg/m^3.)

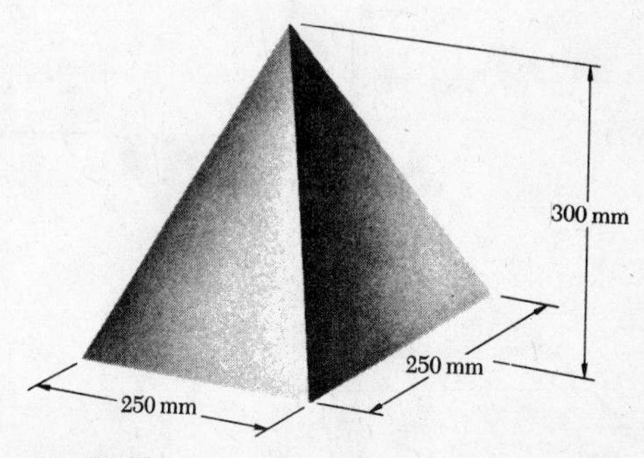

Fig. P5.80

5.81 Locate the centroid of the volume obtained by rotating the area shown about the x axis. The expression obtained may be used to confirm the values given in Fig. 5.21 for the paraboloid (with $n = \frac{1}{2}$) and the cone (with $n = 1$).

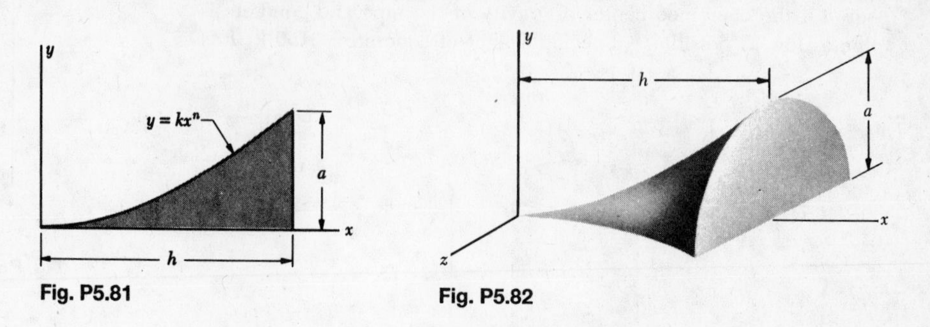

Fig. P5.81

Fig. P5.82

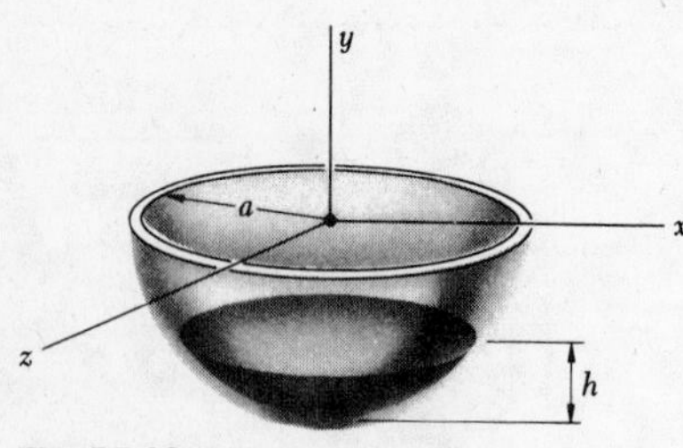

Fig. P5.83

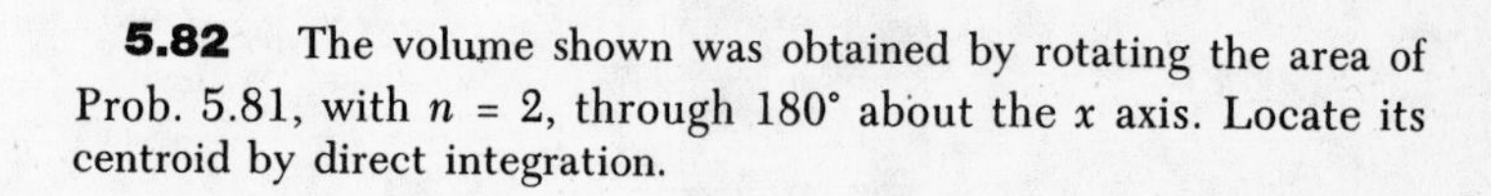

5.82 The volume shown was obtained by rotating the area of Prob. 5.81, with $n = 2$, through 180° about the x axis. Locate its centroid by direct integration.

***5.83** A hemispherical tank of radius a is filled with water to a depth h. Determine by direct integration the center of gravity of the water in the tank.

***5.84** A spherical tank is 2 m in diameter and is filled with water to a depth of 1.5 m. Determine by direct integration the center of gravity of the water in the tank.

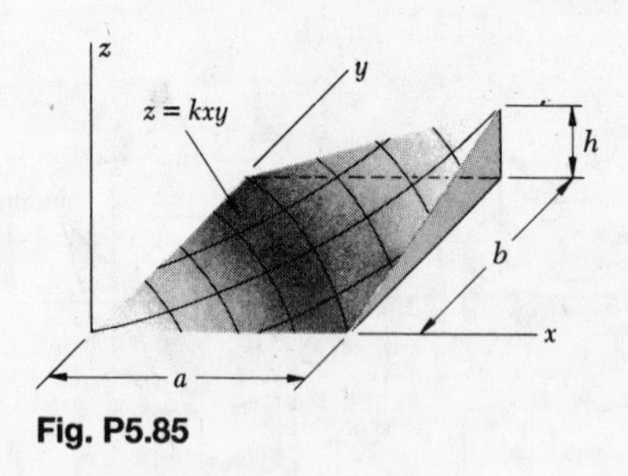

Fig. P5.85

5.85 Determine by direct integration the location of the centroid of the volume between the xy plane and the portion of the hyperbolic paraboloid shown.

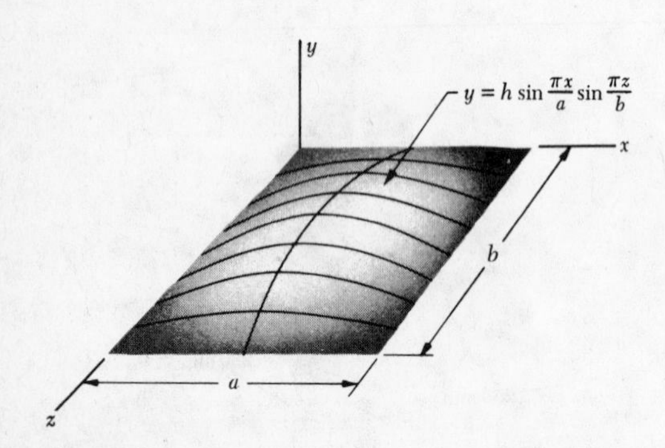

Fig. P5.86

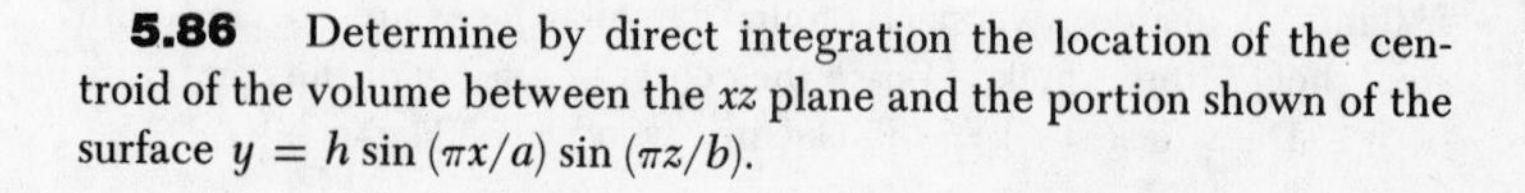

5.86 Determine by direct integration the location of the centroid of the volume between the xz plane and the portion shown of the surface $y = h \sin(\pi x/a) \sin(\pi z/b)$.

CHAPTER 6
ANALYSIS OF STRUCTURES

SECTIONS 6.1 to 6.6

6.1 through 6.6 Using the method of joints, determine the force in each member of the truss shown. State whether each member is in tension or compression.

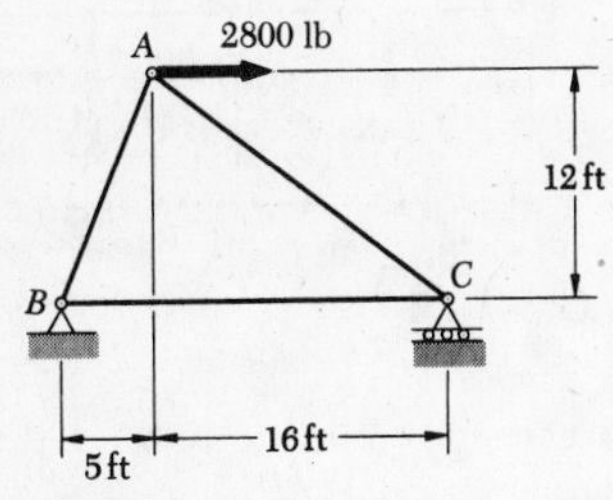

Fig. P6.1

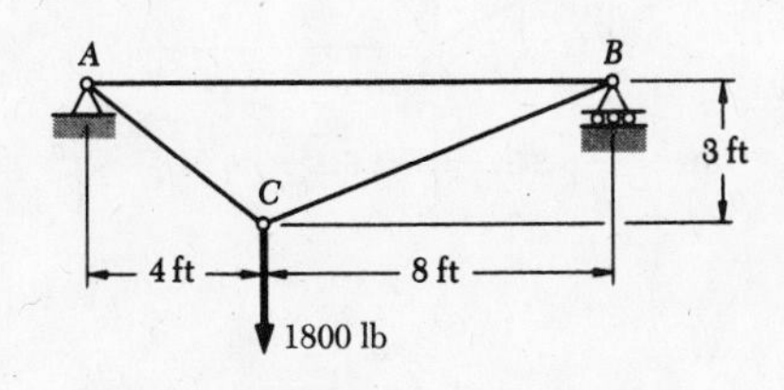

Fig. P6.2

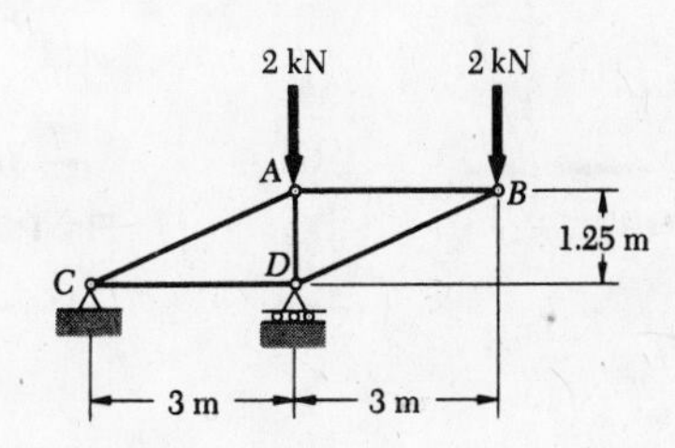

Fig. P6.3

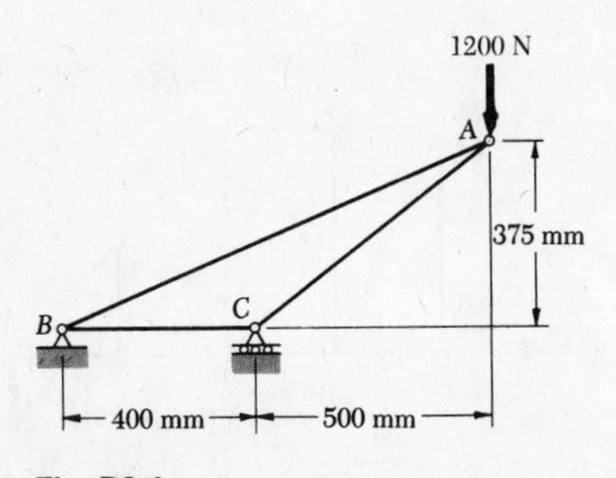

Fig. P6.4

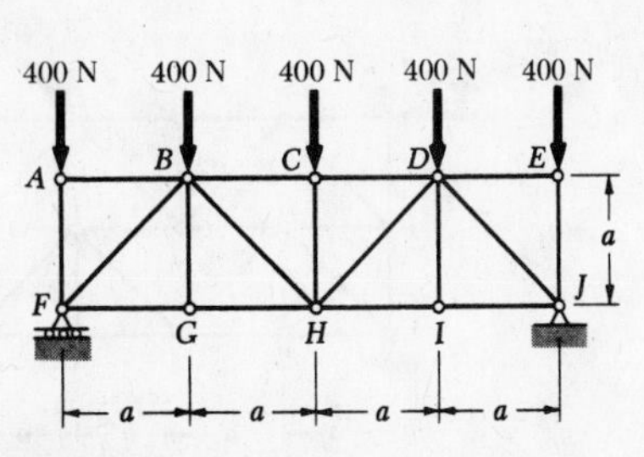

Fig. P6.5

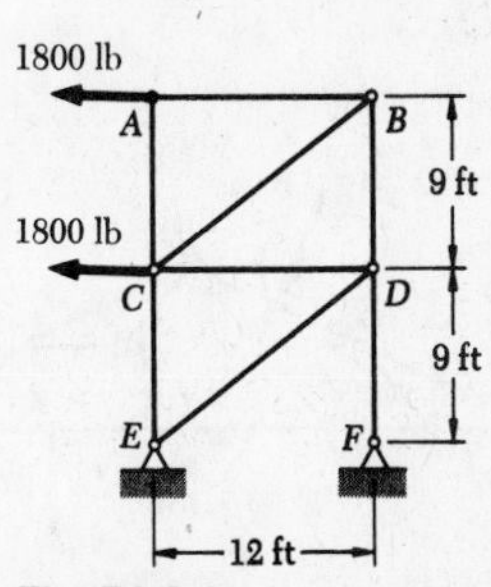

Fig. P6.6

6.7 through 6.12 Using the method of joints, determine the force in each member of the truss shown. State whether each member is in tension or compression.

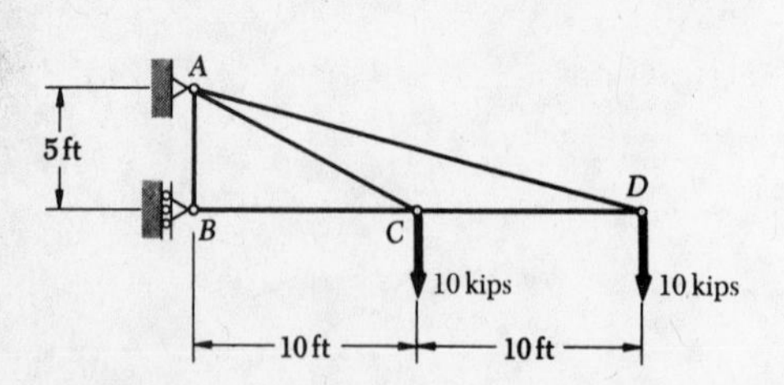

Fig. P6.7

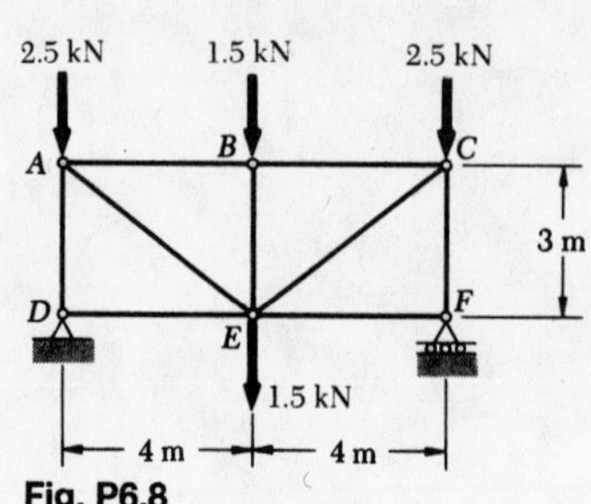

Fig. P6.8

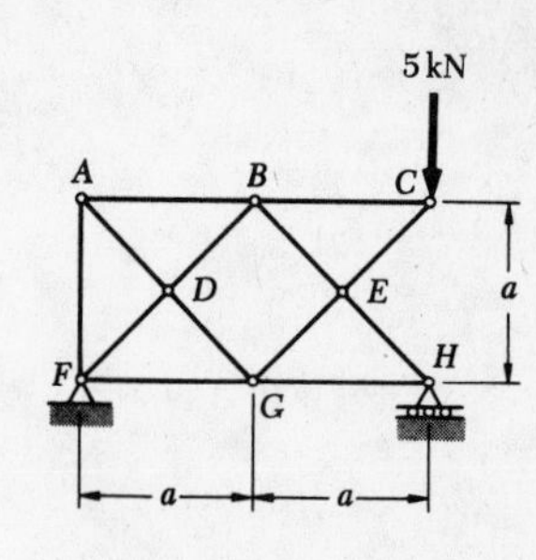

Fig. P6.9

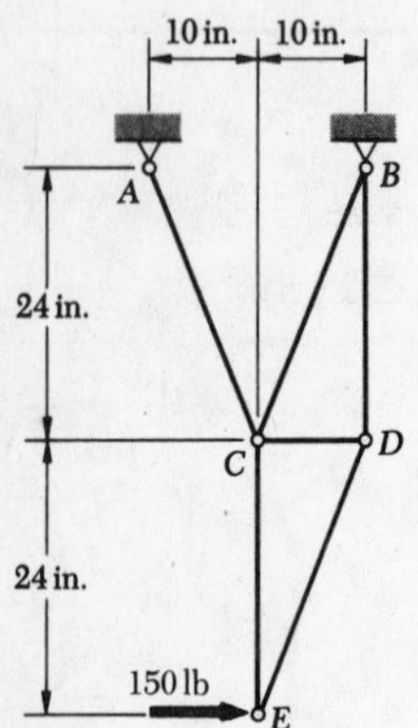

Fig. P6.10

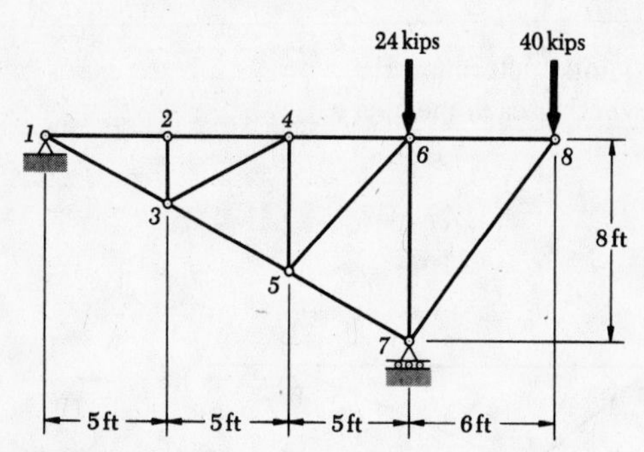

Fig. P6.11

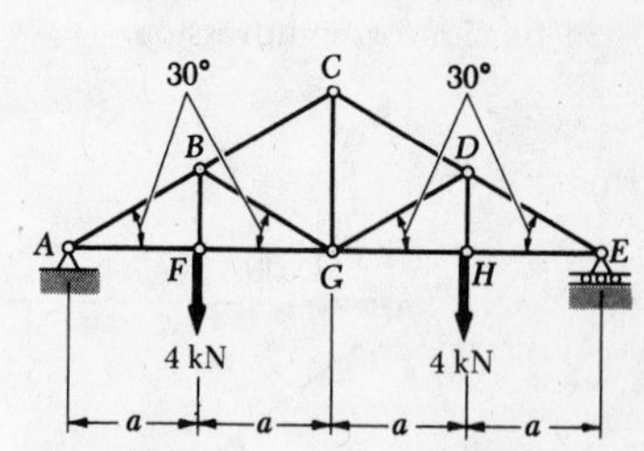

Fig. P6.12

6.13 Determine whether the trusses given in Probs. 6.7, 6.9, 6.14, 6.15, and 6.16 are simple trusses.

6.14 through 6.16 Determine the zero-force members in the truss shown for the given loading.

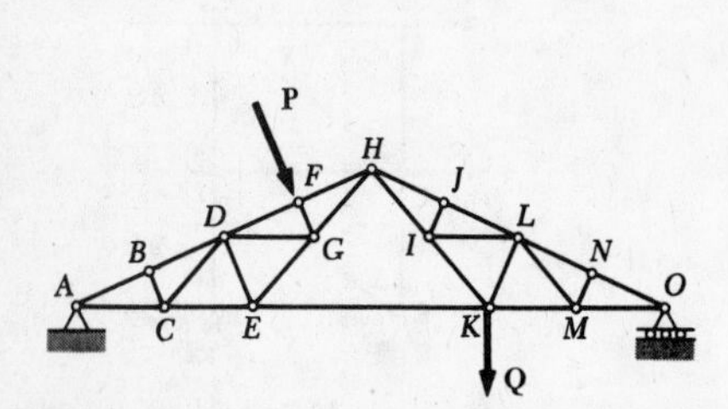

Fig. P6.14

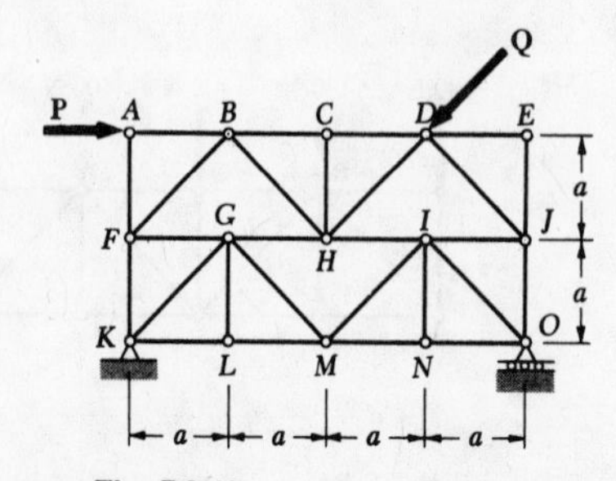

Fig. P6.15

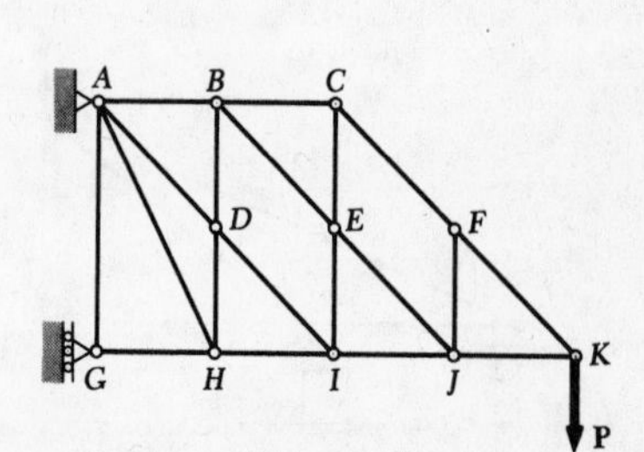

Fig. P6.16

SECTIONS 6.7 and 6.8

6.17 Determine the force in members *BD* and *CD* of the truss shown.

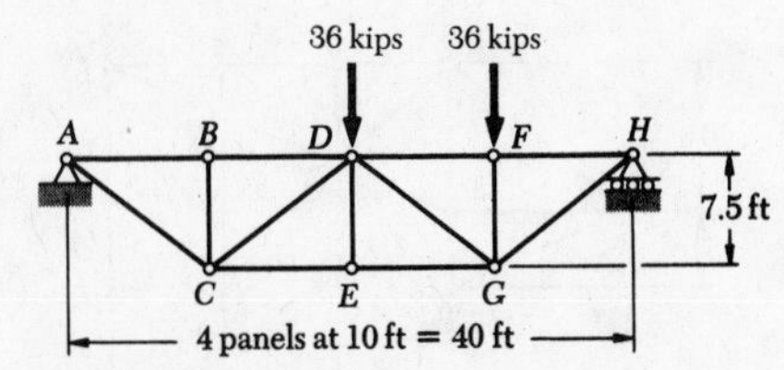

Fig. P6.17 and P6.18

6.18 Determine the force in members *DF* and *DG* of the truss shown.

6.19 Determine the force in members *DF* and *DE* of the truss shown.

6.20 Determine the force in members *CD* and *CE* of the truss shown.

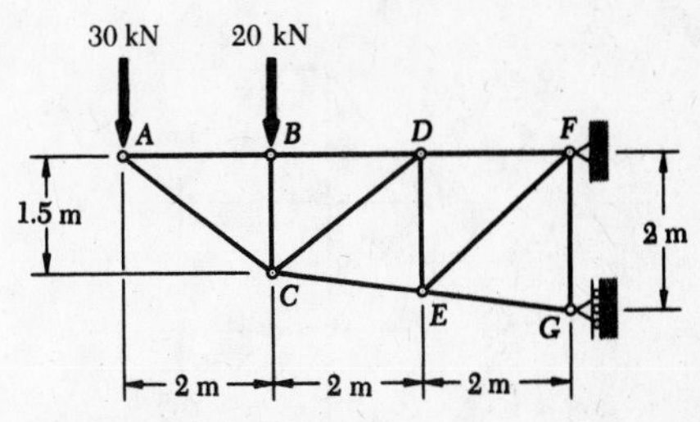

Fig. P6.19 and P6.20

6.21 Determine the force in members *FH*, *GH*, and *GI* of the stadium truss shown.

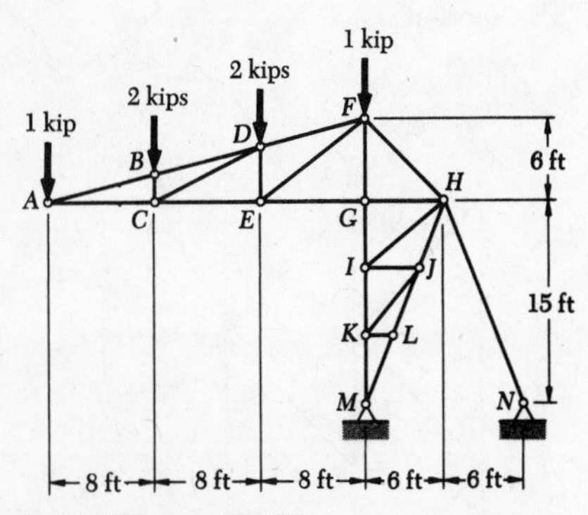

Fig. P6.21 and P6.22

6.22 Determine the force in members *DF*, *DE*, and *CE* of the stadium truss shown.

6.23 Determine the force in members *CE*, *CD*, and *BD* of the truss shown.

6.24 Determine the force in members *EG*, *EF*, and *DF* of the truss shown.

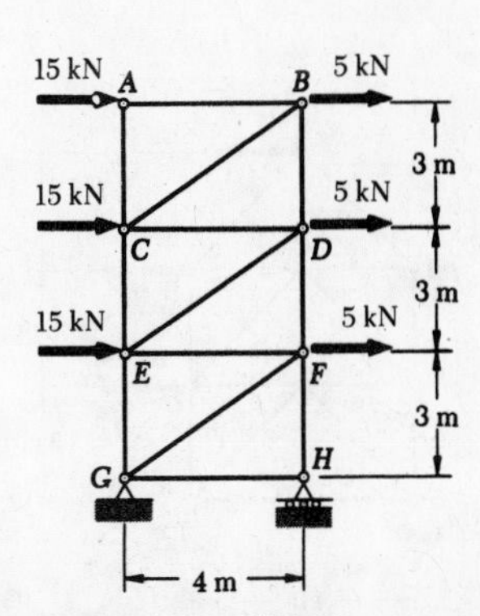

Fig. P6.23 and P6.24

6.25 Determine the force in members EF, FG, and GI of the truss shown.

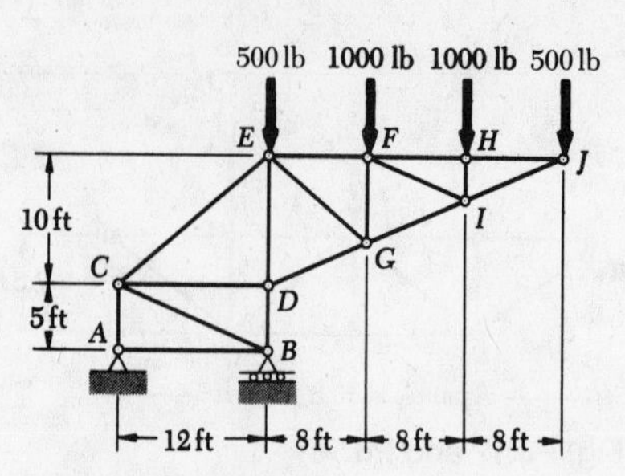

Fig. P6.25 and P6.26

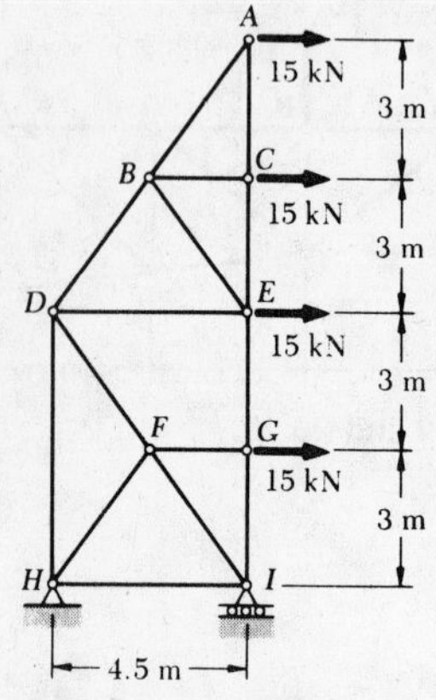

Fig. P6.27 and P6.28

6.26 Determine the force in members CE, CD, and CB of the truss shown.

6.27 Determine the force in members BD and DE of the truss shown.

6.28 Determine the force in members FH and DH of the truss shown.

6.29 Determine the force in members AB and KL of the truss shown. (*Hint.* Use section *a-a.*)

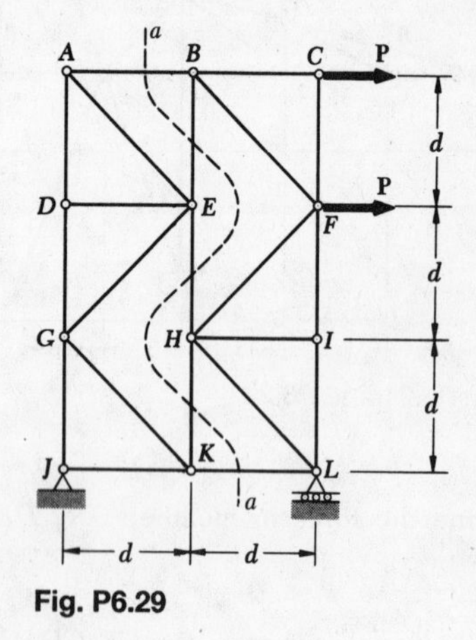

Fig. P6.29

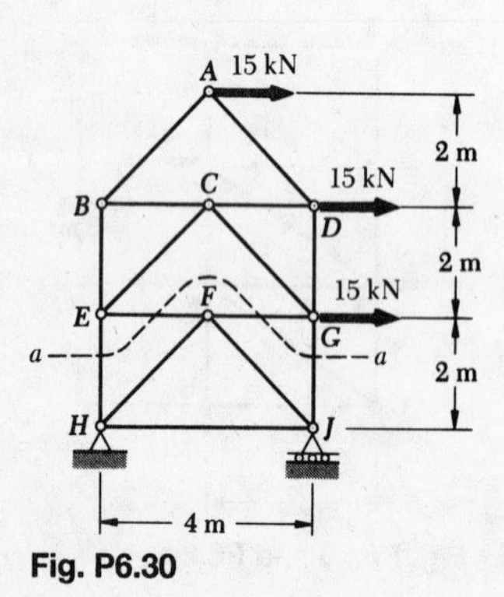

Fig. P6.30

6.30 Determine the force in member GJ of the truss shown. (*Hint.* Use section *a-a.*)

6.31 Determine the force in members AB and EJ of the truss shown, if $\alpha = 0°$. (*Hint.* Use portion IBE of the truss as a free body and apply to joints C, F, G, and H the results obtained in Sec. 6.5.)

6.32 Solve Prob. 6.31, assuming that $\alpha = 90°$.

6.33 The diagonal members in the center panel of the truss shown are very slender and can act only in tension; such members are known as *counters.* Determine the force in members BD and CE and in the counter which is acting under the given loading.

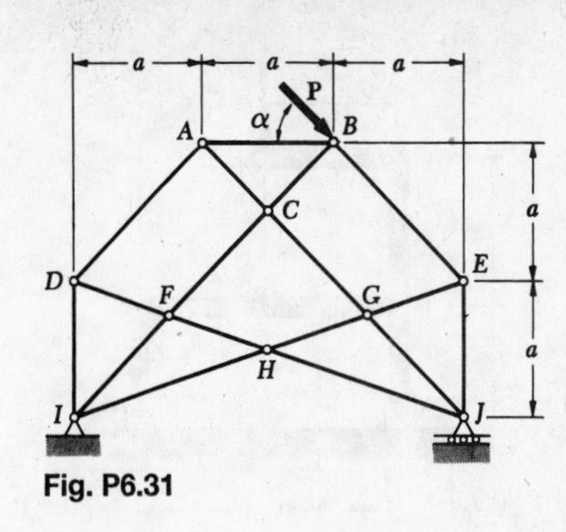

Fig. P6.31

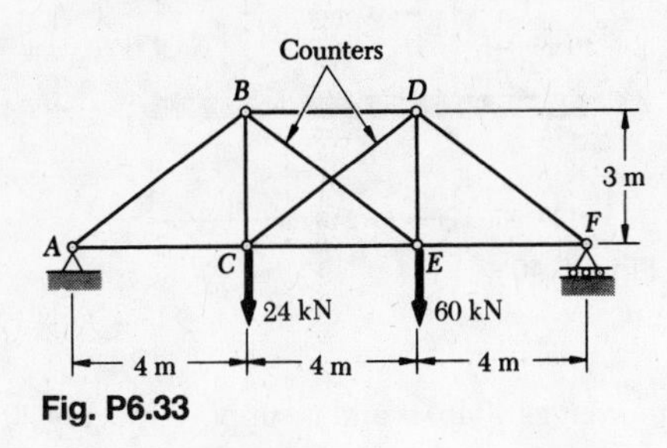

Fig. P6.33

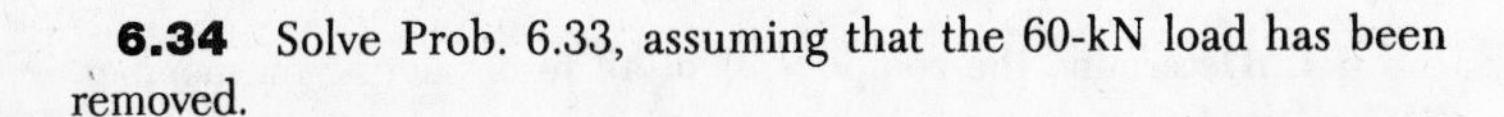

6.34 Solve Prob. 6.33, assuming that the 60-kN load has been removed.

6.35 Determine the force in members BE and CG and in the counter which is acting under the given loading. (See Prob. 6.33 for the definition of a counter.)

6.36 Solve Prob. 6.35, assuming that the 15-kip load has been removed.

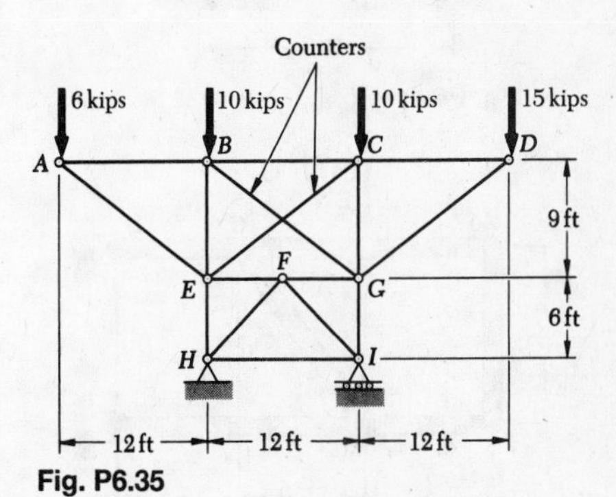

Fig. P6.35

SECTIONS 6.9 to 6.11

6.37 and 6.38 Determine the force in member BD and the components of the reaction at C.

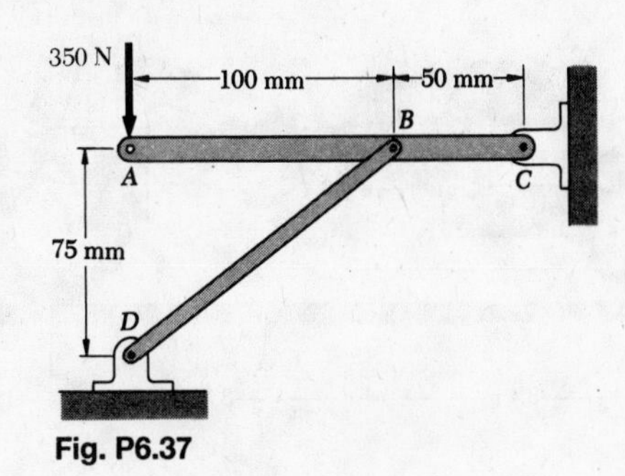

Fig. P6.37

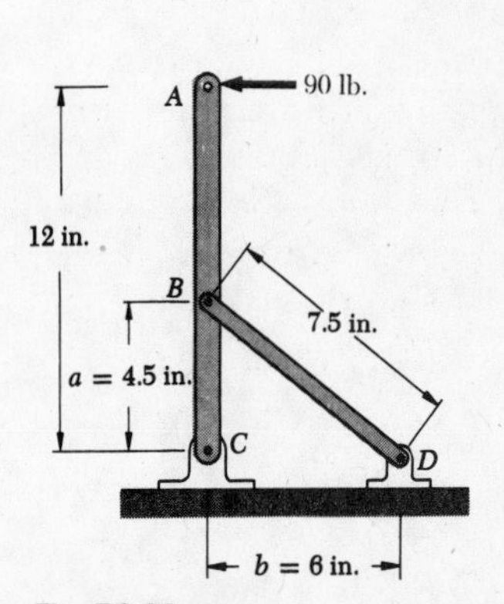

Fig. P6.38

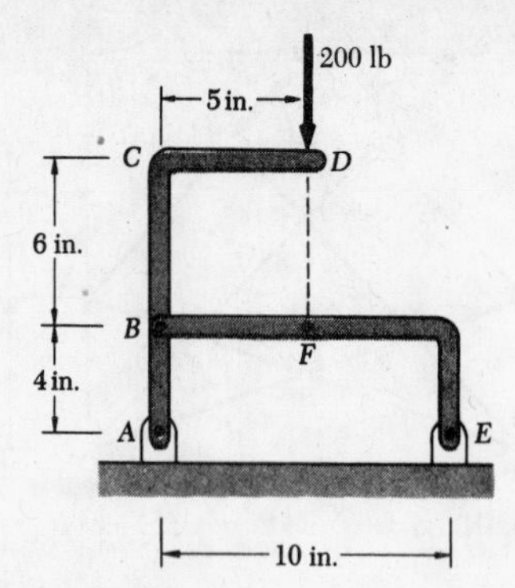

Fig. P6.39

6.39 Determine the components of the force exerted at B on member BE (*a*) if the 200-lb load is applied as shown, (*b*) if the 200-lb load is moved along its line of action and is applied at point F.

6.40 Determine the forces exerted on member AB if the frame is loaded by a clockwise couple of magnitude 6 N · m applied (*a*) at point D, (*b*) at point E. (*c*) Determine the forces exerted on member AB if the frame is loaded by vertical forces applied at D and E which are equivalent to a 6-N · m clockwise couple.

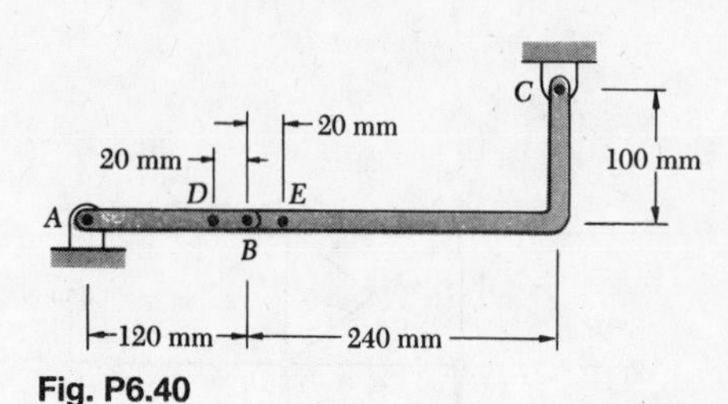

Fig. P6.40

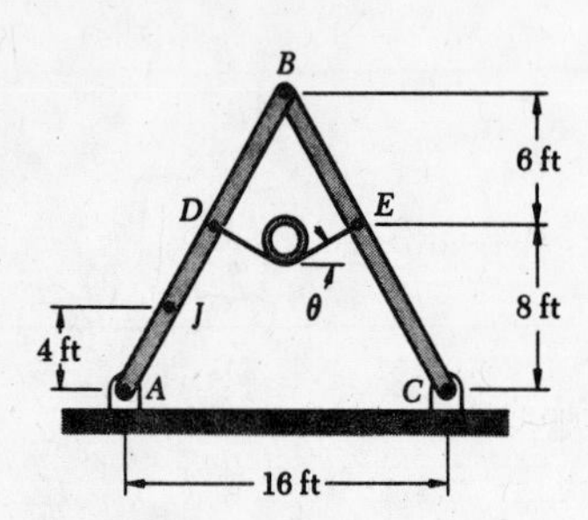

Fig. P6.41

6.41 A pipe weighs 40 lb/ft and is supported every 30 ft by a small frame; a typical frame is shown. Knowing that $\theta = 30°$, determine the components of the reactions and the components of the force exerted at B on member AB.

6.42 Determine the components of all forces acting on member EFG of the frame shown.

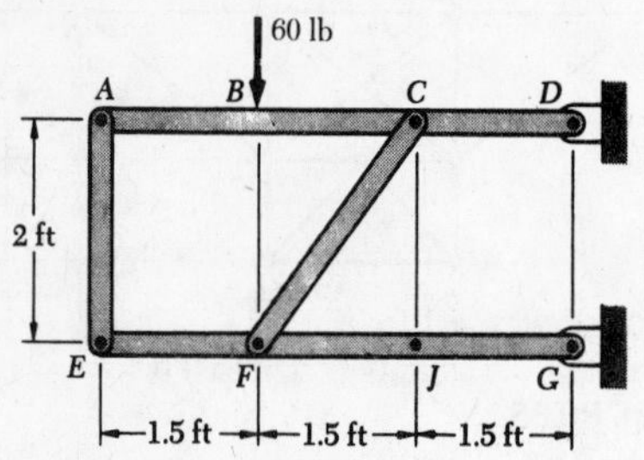

Fig. P6.42

6.43 The tractor and scraper units shown are connected by a vertical pin located 0.6 m behind the tractor wheels. The distance from C to D is 0.75 m. The center of gravity of the 10-Mg tractor unit is located at G_t. The scraper unit and load have a total mass of 50 Mg and a combined center of gravity located at G_s. Knowing that the machine is at rest, with its brakes released, determine (*a*) the reactions at each of the four wheels, (*b*) the forces exerted on the tractor unit at C and D.

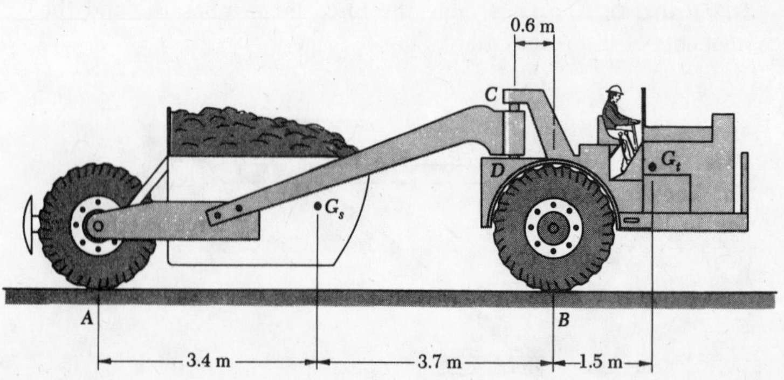

Fig. P6.43

6.44 Determine the components of all forces acting on member AE of the frame shown.

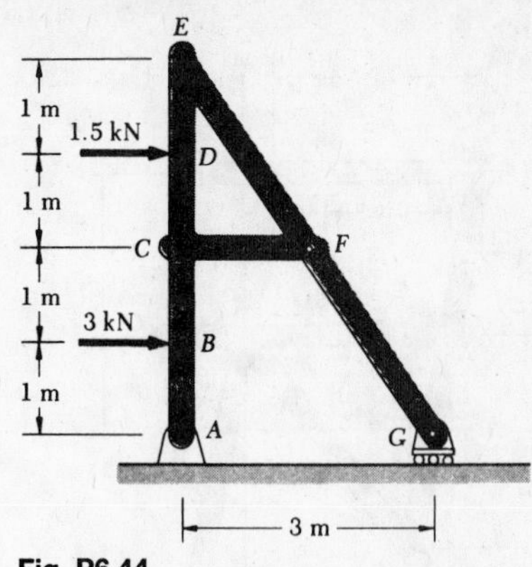

Fig. P6.44

6.45 The axis of the three-hinged arch ABC is a parabola with vertex at B. Knowing that $P = 20$ kN and $Q = 10$ kN, determine (*a*) the components of the reaction at C, (*b*) the components of the force exerted at B on segment AB.

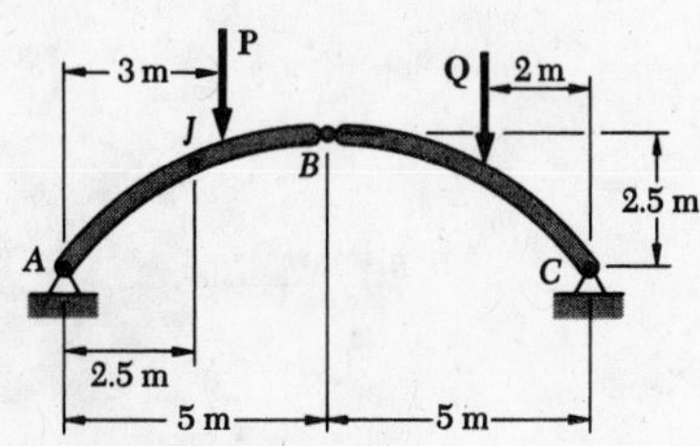

Fig. P6.45 and P6.46

6.46 The axis of the three-hinged arch ABC is a parabola with vertex at B. Knowing that $P = 20$ kN and $Q = 0$, determine (*a*) the components of the reaction at C, (*b*) the components of the force exerted at B on segment AB.

6.47 A vertical load **P** of magnitude 300 lb is applied to member AB. Member AB is placed between two smooth walls and is pin-connected at C to a link CD. Determine all forces exerted on member AB.

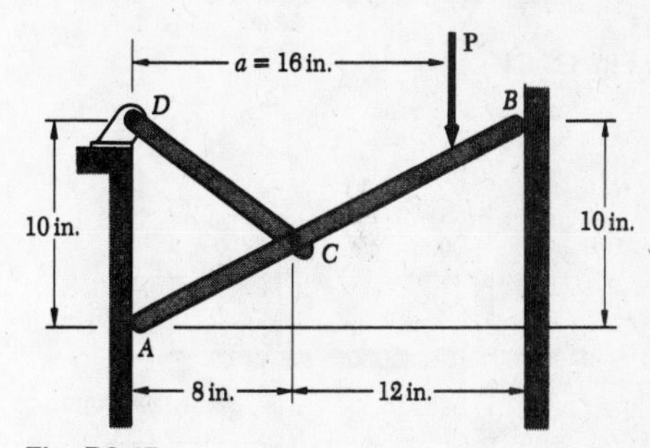

Fig. P6.47

6.48 Determine the reaction at F and the force in members AE and BD.

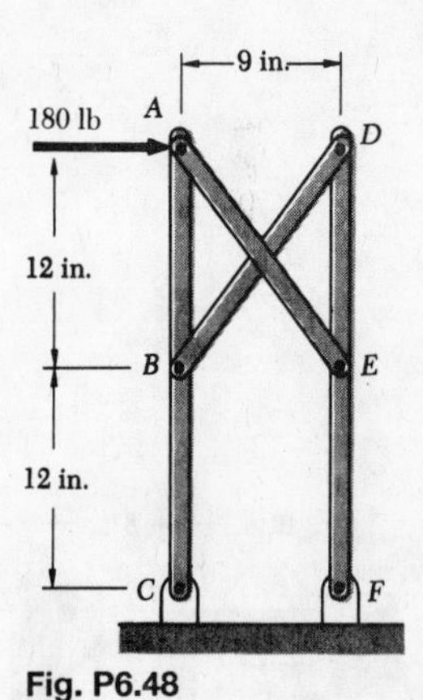

Fig. P6.48

6.49 The frame shown consists of two members ABC and DEF connected by the links AE and BF. Determine the reactions at C and D and the force in links AE and BF.

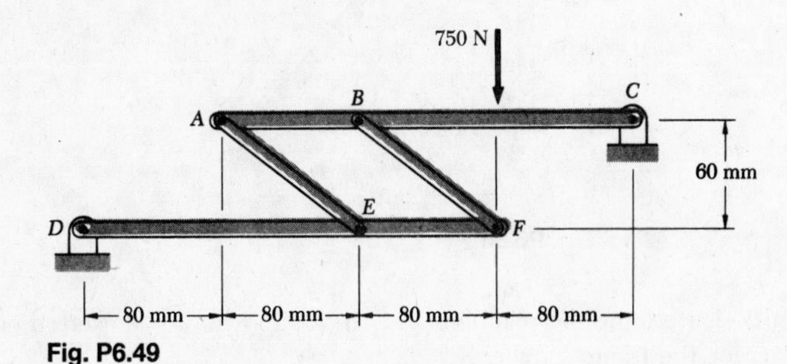

Fig. P6.49

6.50 Solve Prob. 6.49, assuming that the 750-N load is applied at F.

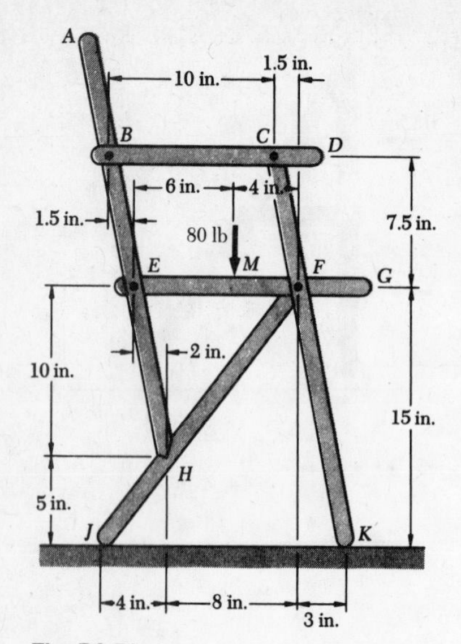

Fig. P6.51

6.51 In the folding chair shown, members *ABEH* and *CFK* are parallel. Determine the components of all forces acting on member *ABEH* when a 160-lb person sits in the chair. It may be assumed that the floor is frictionless and that half the person's weight is carried by each side of the chair and is applied at point *M*.

6.52 The rigid bar *EFG* is supported by the truss system shown. Determine the force in each of the two-force members.

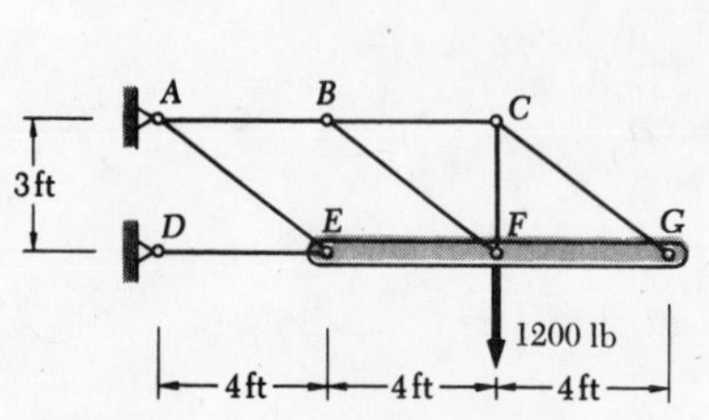

Fig. P6.52

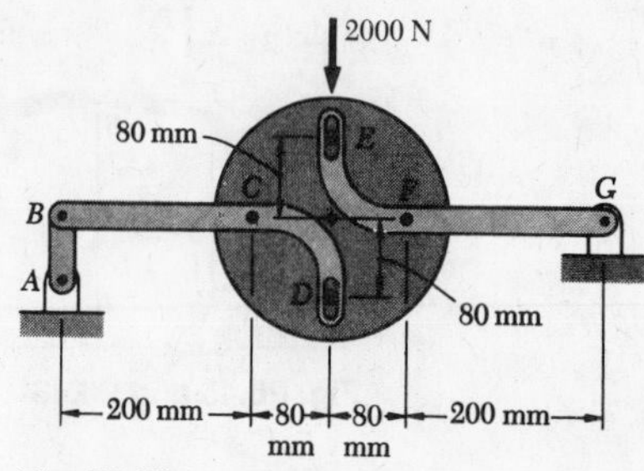

Fig. P6.53

6.53 Two arms *BCD* and *EFG* are connected to a 200-mm-diameter disk by four pins which are attached to the disk. Assuming that the pins at *D* and *E* may slide in the vertical slots, determine the components of all forces exerted on the disk when a 2000-N load is applied to the disk as shown.

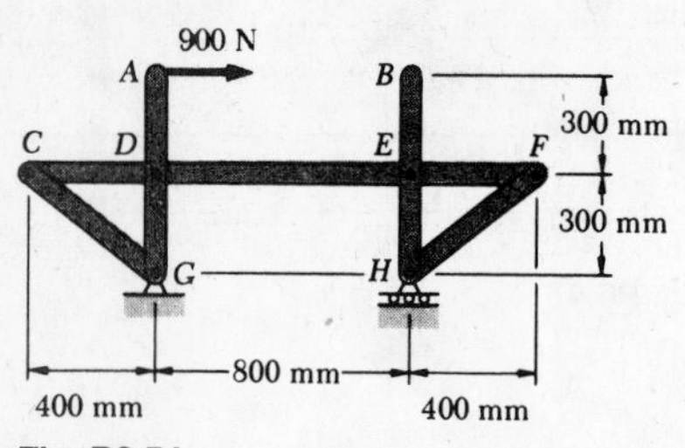

Fig. P6.54

6.54 Determine the components of all forces acting on member *CDEF*.

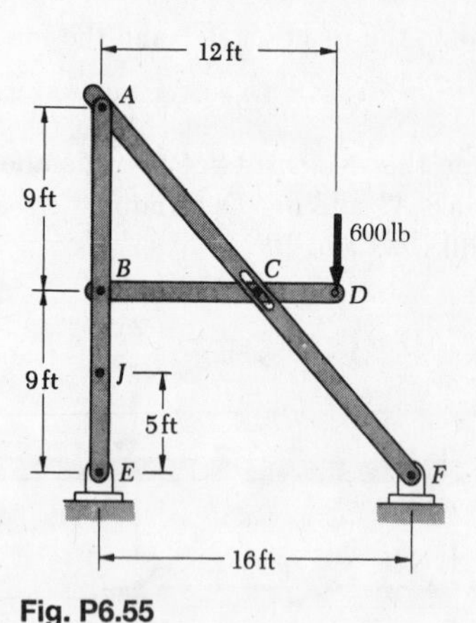

Fig. P6.55

6.55 Determine the reactions at *E* and *F* and the force exerted on pin *C* for the frame shown.

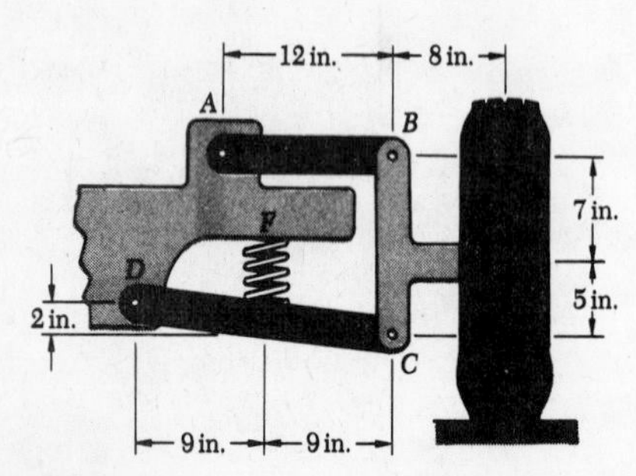

Fig. P6.56

6.56 An automobile front-wheel assembly supports 750 lb. Determine the force exerted by the spring and the components of the forces exerted on the frame at points *A* and *D*.

SECTION 6.12

6.57 A 360-N force is applied to the toggle vise at C. Determine (a) the horizontal force exerted on the block at D, (b) the force exerted on member ABC at B.

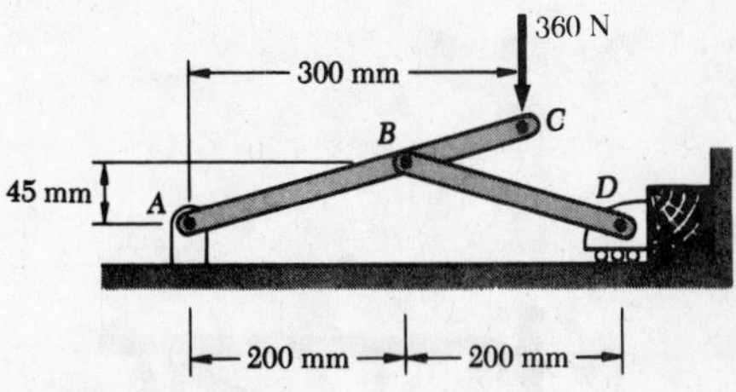

Fig. P6.57

6.58 Two 300-N forces are applied to the handles of the pliers as shown. Determine (a) the magnitude of the forces exerted on the rod, (b) the force exerted by the pin at A on portion AB of the pliers.

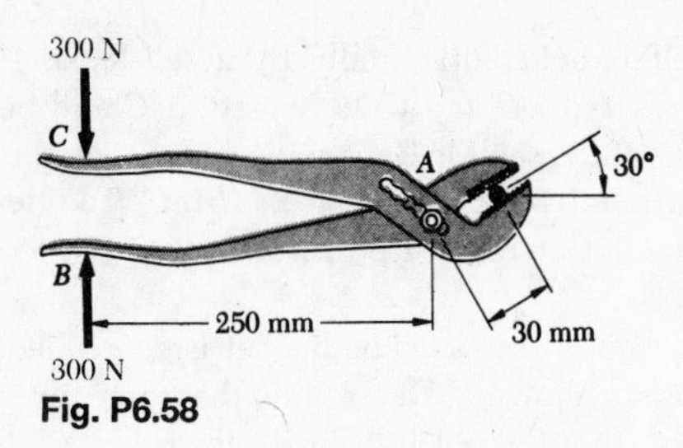

Fig. P6.58

6.59 A cylinder weighs 400 lb and is lifted by a pair of tongs as shown. Determine the forces exerted at D and C on the tong BCD.

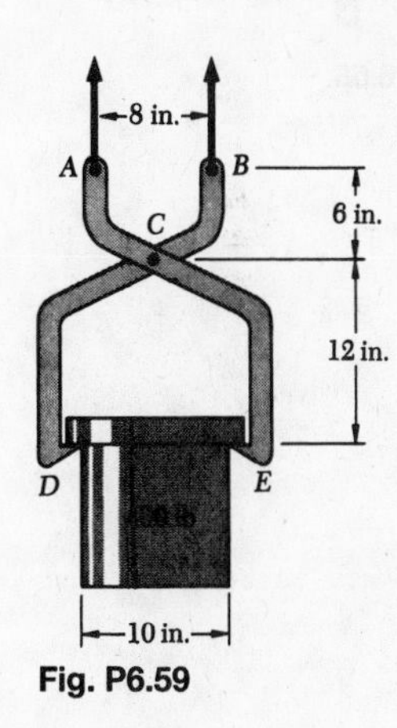

Fig. P6.59

6.60 In using the boltcutter shown, a worker applies two 100-lb forces to the handles. Determine the magnitude of the forces exerted by the cutter on the bolt.

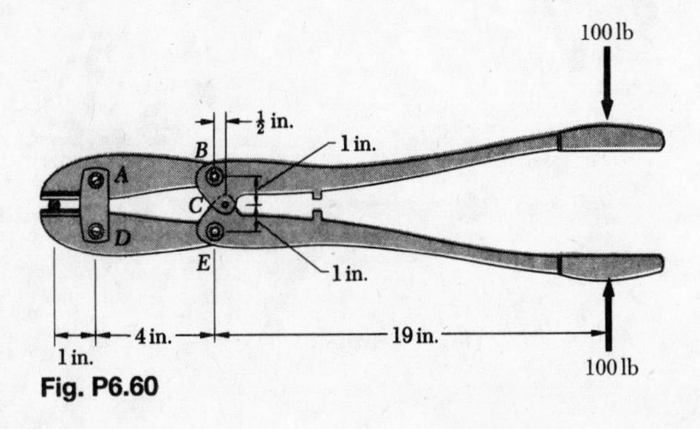

Fig. P6.60

6.61 The retractable shelf is maintained in the position shown by two identical linkage-and-spring systems; only one of the systems is shown. A 20-kg machine is placed on the shelf so that half of its weight is supported by the system shown. Determine (a) the force in link AB, (b) the tension in the spring.

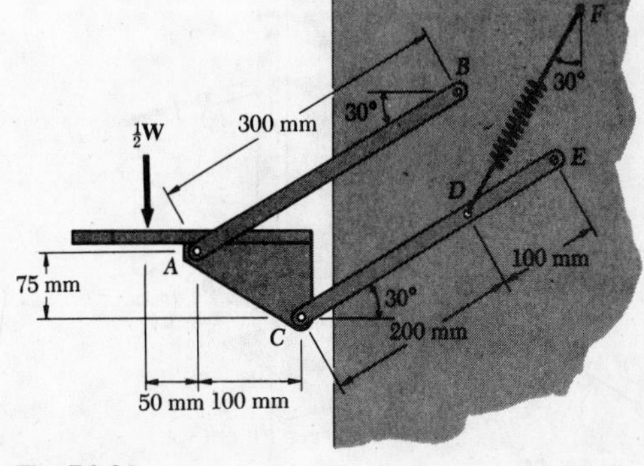

Fig. P6.61

6.62 Determine the couple **M** which must be applied to the crank *CD* to hold the mechanism in equilibrium. The block at *D* is pinned to the crank *CD* and is free to slide in a slot cut in member *AB*.

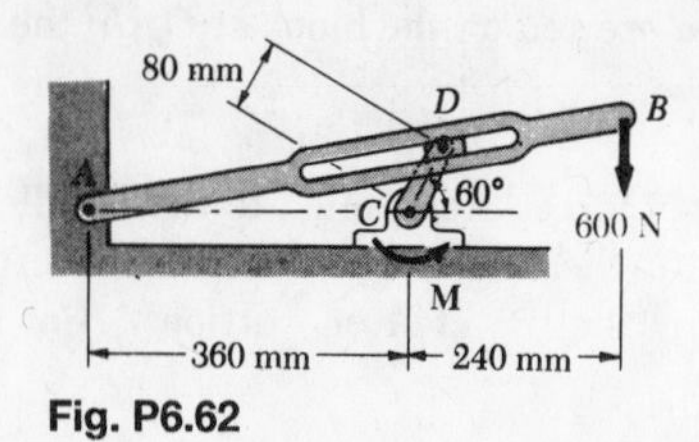

Fig. P6.62

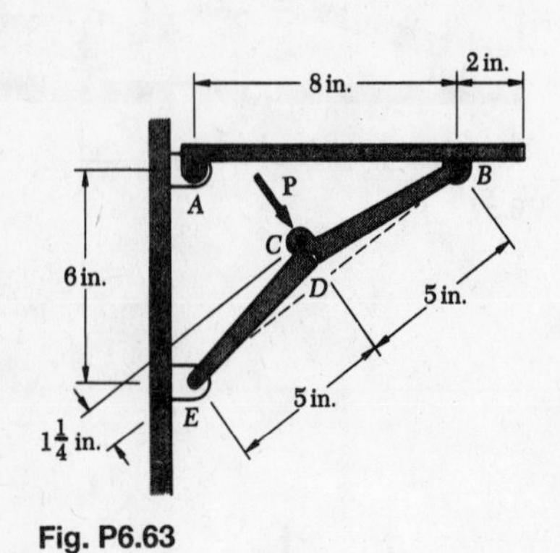

Fig. P6.63

6.63 A shelf is held horizontally by a self-locking brace which consists of two parts *EDC* and *CDB* hinged at *C* and bearing against each other at *D*. If the shelf is 10 in. wide and weighs 24 lb, determine the force *P* required to release the brace. (*Hint.* To release the brace, the forces of contact at *D* must be zero.)

6.64 The action of the backhoe bucket is controlled by the three hydraulic cylinders shown. Determine the force exerted by each cylinder in supporting the 3,000-lb load shown.

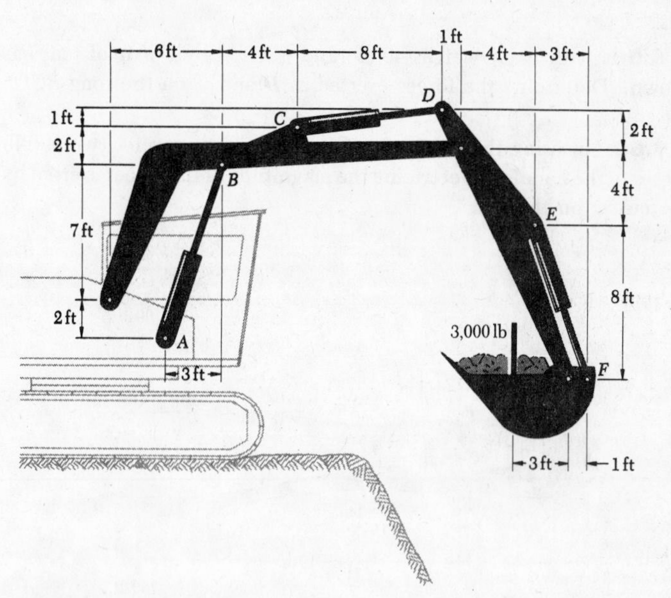

Fig. P6.64

Fig. P6.65

6.65 The four gears are rigidly attached to shafts which are held by frictionless bearings. If $M_1 = 24$ N · m and $M_2 = 0$, determine (*a*) the couple $\mathbf{M}_3$ which must be applied for equilibrium, (*b*) the reactions at *G* and *H*.

6.66 Solve Prob. 6.65 if $M_1 = 48$ N • m and $M_2 = -18$ N • m.

6.67 The total weight of the 1-yd clamshell bucket shown is 4500 lb. The centers of gravity of sections *DEJ* and *EFK*, which weigh 2000 lb each, are located at *G* and *H*, respectively. The double-sheave pulley *E* and a counterweight located at *E* together weigh 400 lb. Determine the tension in cable *1* and cable *2* for the position shown. (Neglect the effect of the horizontal distance between the cables.)

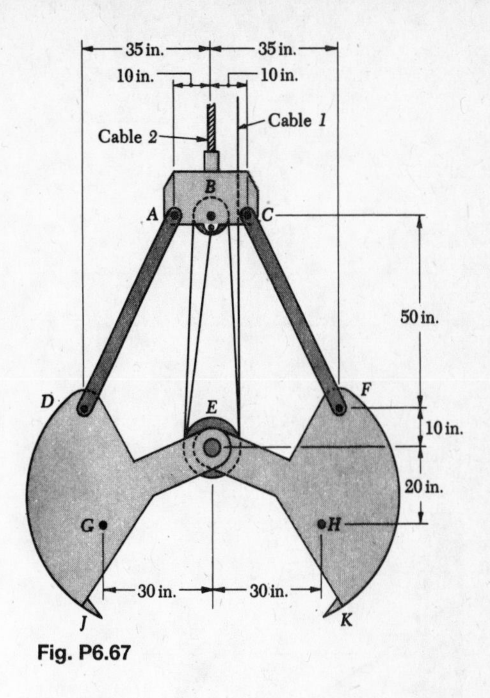

Fig. P6.67

6.68 For the bevel-gear system shown, determine the required value of α if the ratio of M_B to M_A is to be 3.

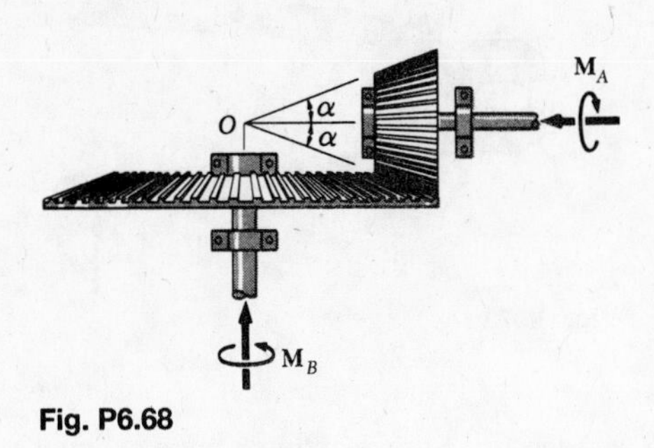

Fig. P6.68

6.69 In the boring rig shown, the center of gravity of the 3000-kg tower is located at point *G*. For the position shown, determine the force exerted by the hydraulic cylinder *AB*.

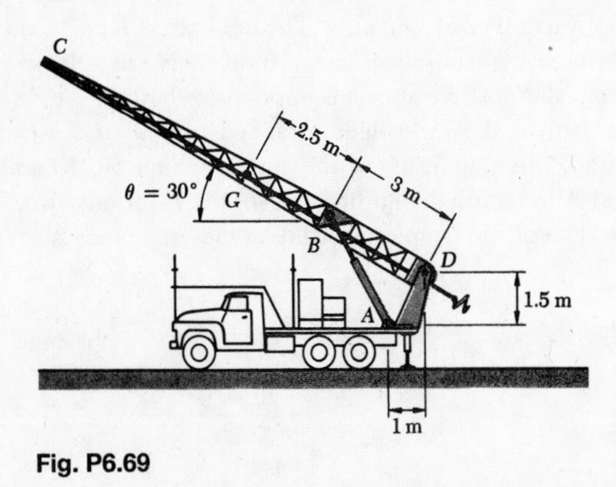

Fig. P6.69

6.70 Solve Prob. 6.69 for the position $\theta = 60°$.

6.71 A force **P** of magnitude 400 lb is applied to the piston of the engine system shown. For each of the two positions shown, determine the couple **M** required to hold the system in equilibrium.

6.72 A couple **M** of magnitude 210 lb · ft is applied to the crank of the engine system shown. For each of the two positions shown, determine the force **P** required to hold the system in equilibrium.

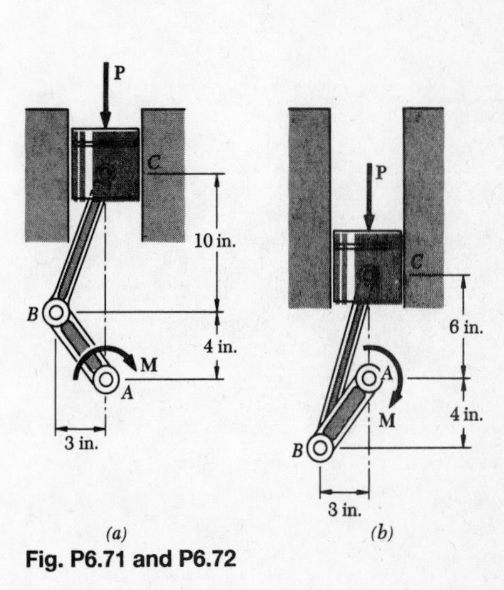

Fig. P6.71 and P6.72

6.73 In the pliers shown, the clamping jaws remain parallel as objects of various sizes are held. If gripping forces of magnitude $Q = 450$ lb are desired, determine the magnitude P of the forces which must be applied. Assume that pins B and E slide freely in the slots cut in the jaws.

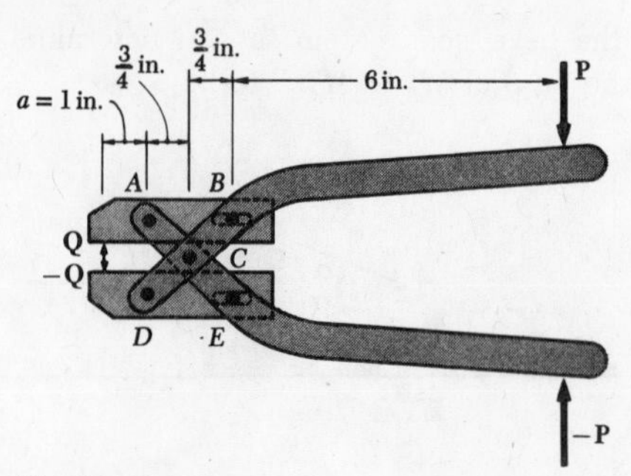

Fig. P6.73

Fig. P6.74

6.74 Two identical linkage-and-hydraulic-cylinder systems control the position of the forks of a fork-lift truck; only one system is shown. Knowing that the load supported by the one system shown is 1500 lb, determine (a) the force exerted by the hydraulic cylinder on point D, (b) the components of the force exerted on member ACE at point C.

***6.75** Two shafts AC and CF, which lie in the vertical xy plane, are connected by a universal joint at C. The bearings at B and D do not exert any axial force. A couple of magnitude 50 N · m (clockwise when viewed from the positive x axis) is applied to shaft CF at F. At a time when the arm of the crosspiece attached to shaft CF is horizontal, determine (a) the magnitude of the couple which must be applied to shaft AC at A to maintain equilibrium, (b) the reactions at B, D, and E. (*Hint.* The sum of the couples exerted on the crosspiece must be zero.)

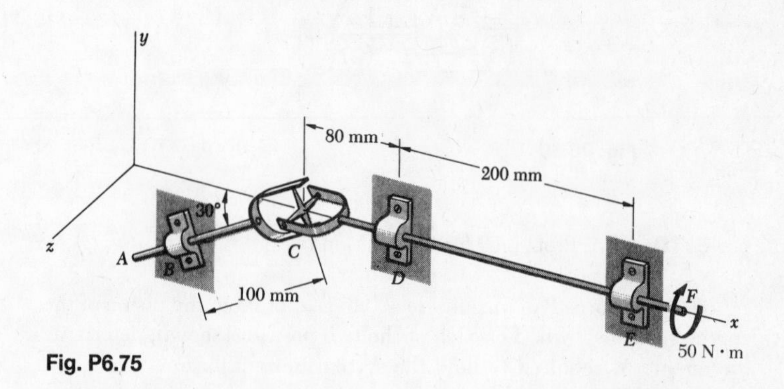

Fig. P6.75

***6.76** Solve Prob. 6.75, assuming that the arm of the crosspiece attached to shaft CF is vertical.

CHAPTER 7
FORCES IN BEAMS AND CABLES

SECTIONS 7.1 AND 7.2

7.1 through 7.4 Determine the internal forces (axial force, shearing force, and bending moment) at point J of the structure indicated:

- **7.1** Frame and loading of Prob. 6.41.
- **7.2** Frame and loading of Prob. 6.42.
- **7.3** Frame and loading of Prob. 6.45.
- **7.4** Frame and loading of Prob. 6.46.

7.5 The bracket AD is supported by a pin at A and by the cable DE. Determine the internal forces just to the left of the load if $a = 200$ mm.

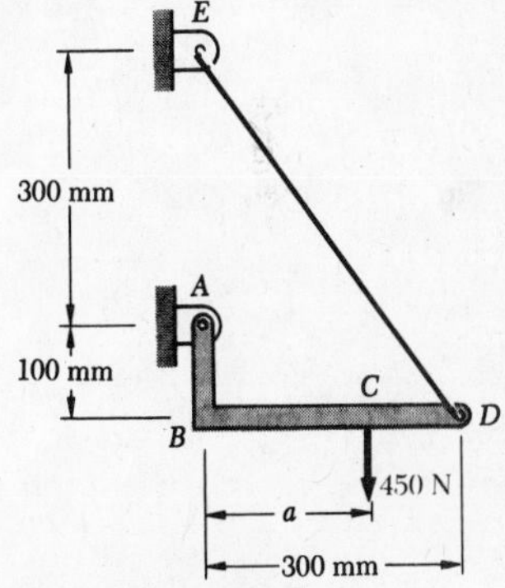

Fig. P7.5 and P7.6

7.6 The bracket AD is supported by a pin at A and by the cable DE. Determine the distance a for which the bending moment at B is equal to the bending moment at C.

7.7 and 7.8 A steel channel forms one side of a flight of stairs. If one channel weighs w lb/ft, determine the internal forces at the center of one channel due to its own weight, (a) in terms of w, L, and θ, (b) if $w = 20$ lb/ft, $l = 12$ ft, and $h = 9$ ft.

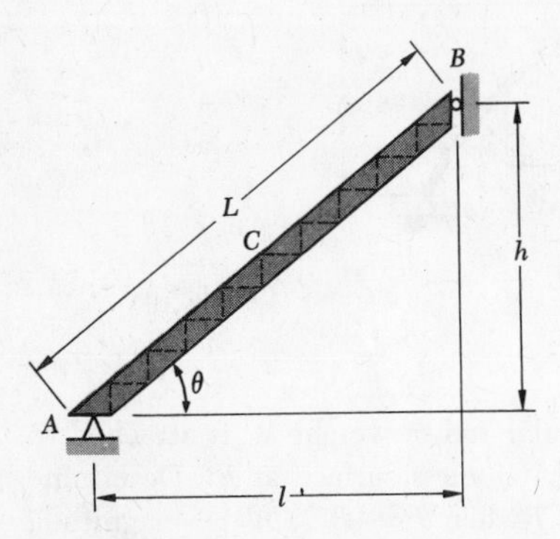

Fig. P7.7

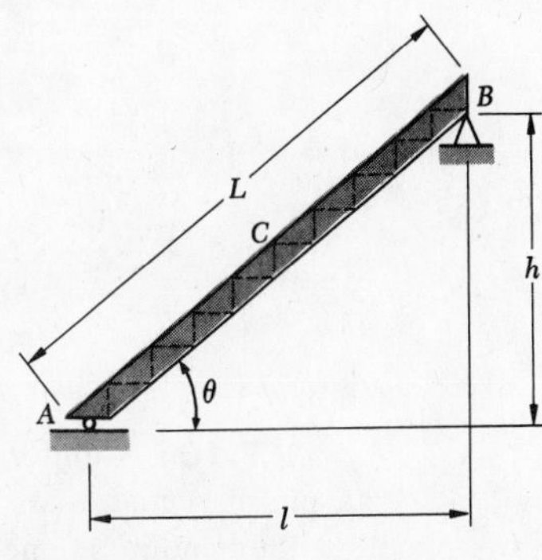

Fig. P7.8

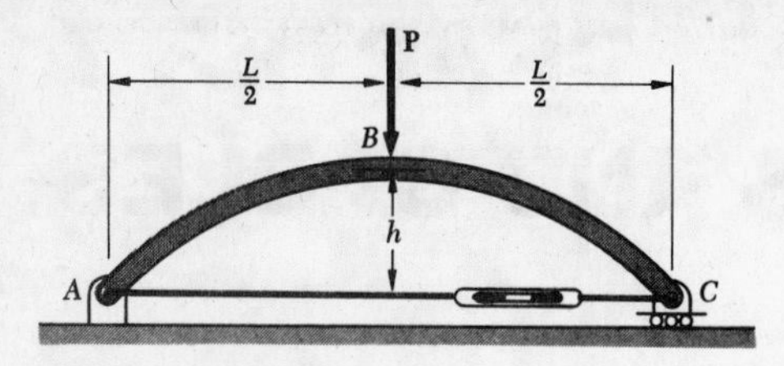

Fig. P7.9 and P7.10

7.9 The axis of the curved member ABC is parabolic. If the tension in the cable AC is 200 lb and the magnitude of the force **P** is 100 lb, determine the internal forces at a distance $\frac{1}{4}L$ from the left end. Assume $h = 4$ in. and $L = 24$ in.

7.10 Denoting by T the tension in the cable AC, determine (*a*) the internal forces just to the left of point B, (*b*) the value of T for which the bending moment at B is zero.

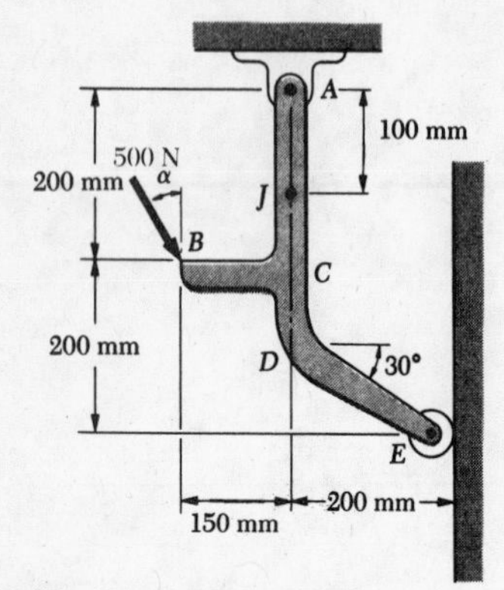

Fig. P7.11 and P7.12

7.11 Determine the internal forces at point J when $\alpha = 90°$.

7.12 Determine the internal forces at point J when $\alpha = 0$.

7.13 Determine the internal forces at point J of the adjustable hanger shown.

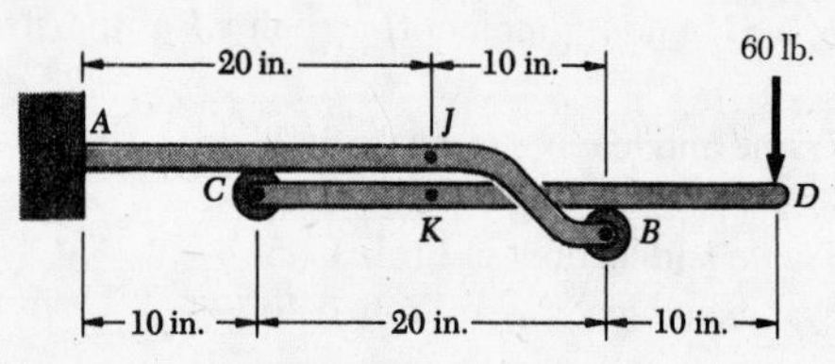

Fig. P7.13 and P7.14

7.14 Determine the internal forces at point K of the adjustable hanger shown.

7.15 The axis of the curved member AB is a parabola with vertex at A. (*a*) Determine the magnitude and location of the maximum bending moment. (*b*) Noting that AB is a two-force member, determine the maximum perpendicular distance from the chord AB to the curved member.

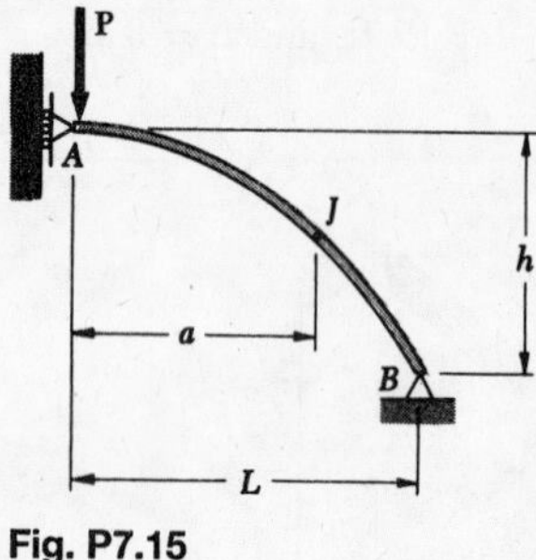

Fig. P7.15

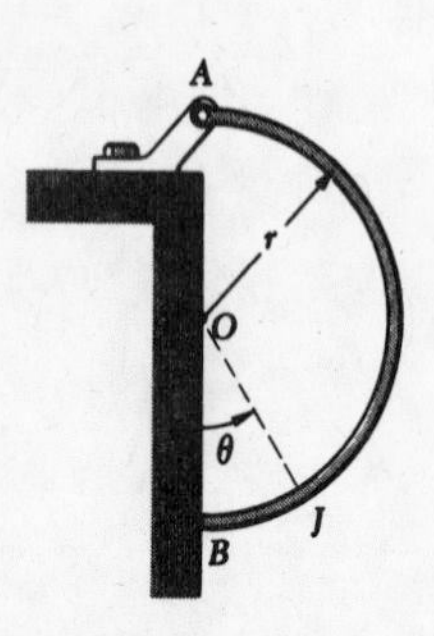

Fig. P7.16

***7.16** A uniform semicircular rod of weight W is attached to a pin at A and bears against a frictionless surface at B. Determine (*a*) the bending moment at point J when $\theta = 90°$, (*b*) the magnitude and location of the maximum bending moment.

SECTIONS 7.3 to 7.5

7.17 through 7.20 Draw the shear and bending-moment diagrams for the beam and loading shown.

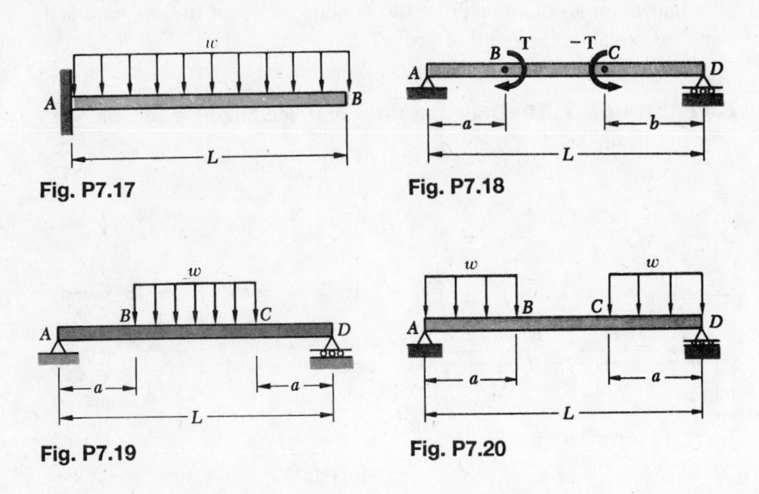

Fig. P7.17 Fig. P7.18

Fig. P7.19 Fig. P7.20

7.21 and 7.22 Draw the shear and bending-moment diagrams for the beam *AB*.

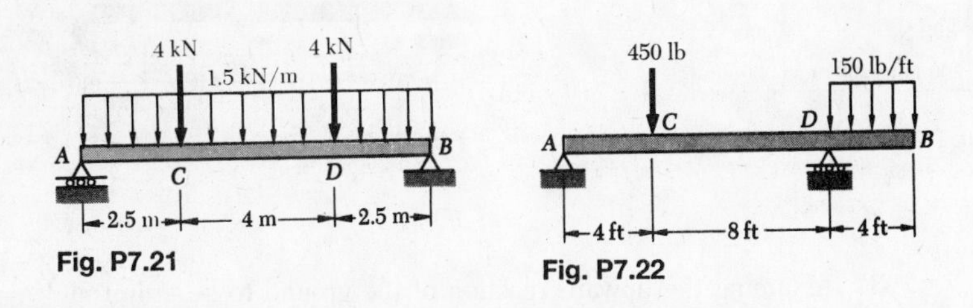

Fig. P7.21 Fig. P7.22

7.23 Draw the shear and bending-moment diagrams for the beam *AB* if $a = 6$ ft.

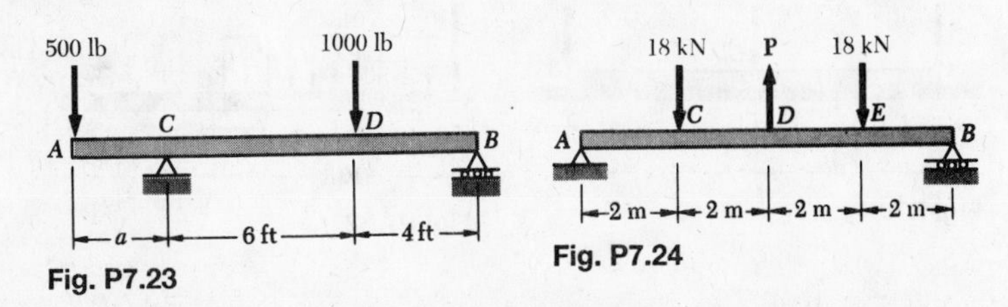

Fig. P7.23 Fig. P7.24

7.24 Draw the shear and bending-moment diagrams for the beam *AB* if the magnitude of the upward force **P** is 4 kN.

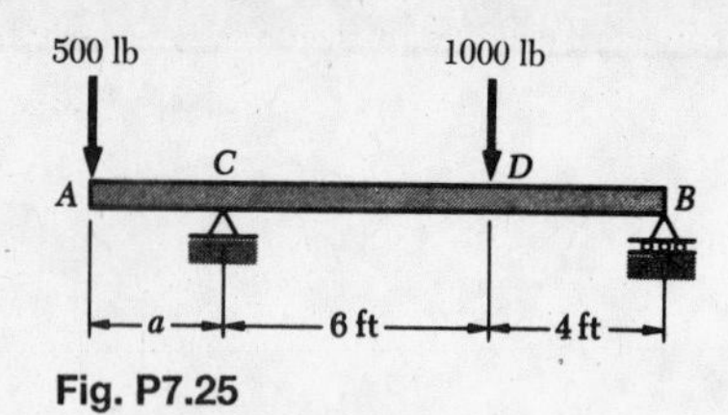

Fig. P7.25

***7.25** Determine the distance a for which the maximum absolute value of the bending moment in the beam is as small as possible. (*Hint.* Draw the bending-moment diagram and then equate the maximum positive and negative bending moments obtained.)

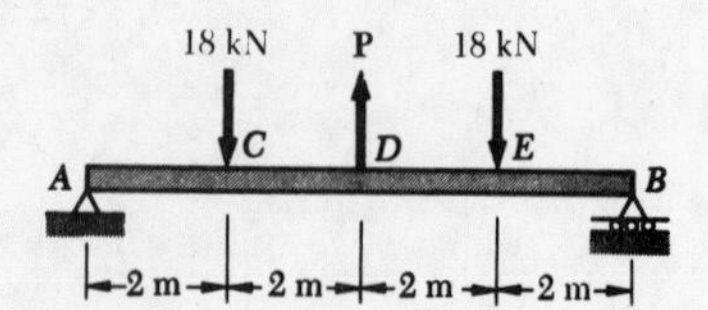

Fig. P7.26

***7.26** Determine the magnitude of the upward force **P** for which the maximum absolute value of the bending moment in the beam is as small as possible. (See hint of Prob. 7.25.)

7.27 through 7.30 Draw the shear and bending-moment diagrams for the beam AB.

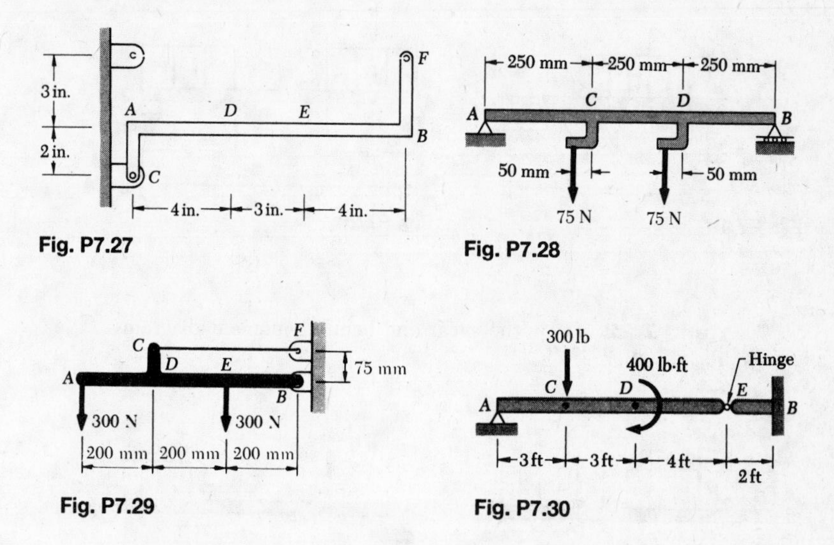

Fig. P7.27

Fig. P7.28

Fig. P7.29

Fig. P7.30

7.31 Assuming the upward reaction of the ground to be uniformly distributed, draw the shear and bending-moment diagrams for the beam AB.

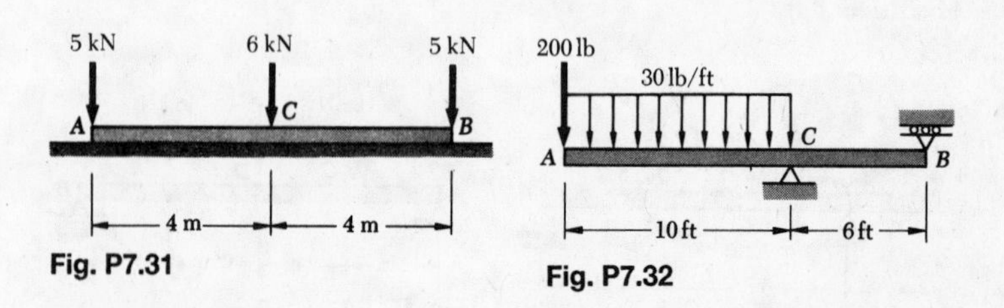

Fig. P7.31

Fig. P7.32

7.32 Draw the shear and bending-moment diagrams for the beam AB.

SECTION 7.6

7.33 Using the methods of Sec. 7.6, solve Prob. 7.17.
7.34 Using the methods of Sec. 7.6, solve Prob. 7.31.
7.35 Using the methods of Sec. 7.6, solve Prob. 7.19.
7.36 Using the methods of Sec. 7.6, solve Prob. 7.20.
7.37 Using the methods of Sec. 7.6, solve Prob. 7.21.
7.38 Using the methods of Sec. 7.6, solve Prob. 7.22.
7.39 Using the methods of Sec. 7.6, solve Prob. 7.23.
7.40 Using the methods of Sec. 7.6, solve Prob. 7.24.

7.41 through 7.44 Draw the shear and bending-moment diagrams for the beam and loading shown.

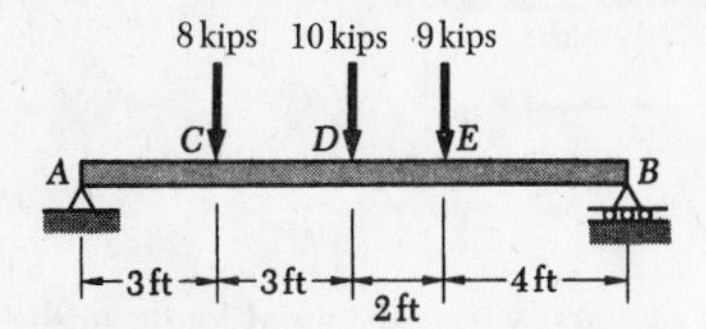

Fig. P7.41

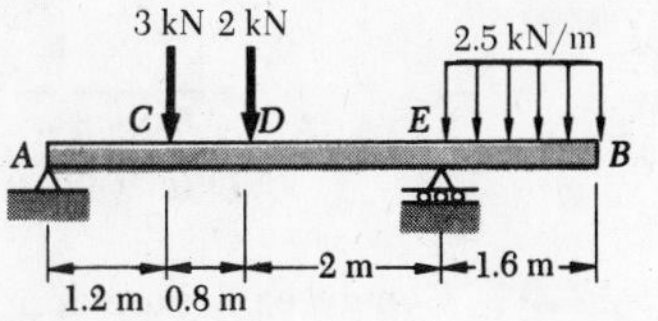

Fig. P7.42

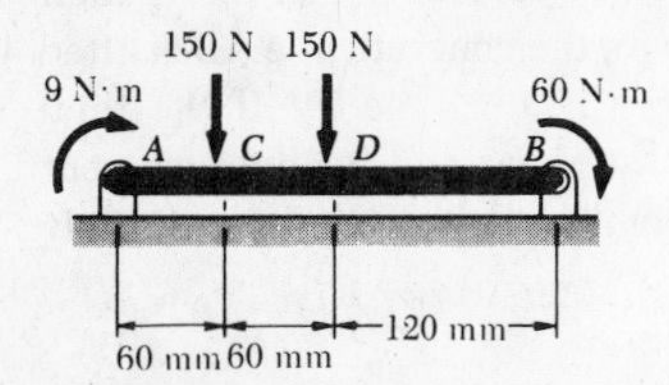

Fig. P7.43

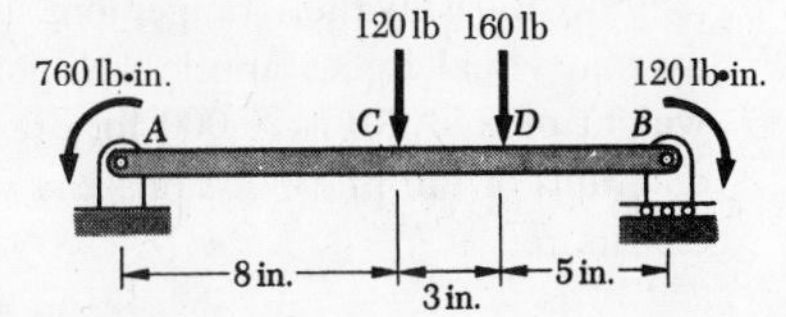

Fig. P7.44

7.45 through 7.48 Draw the shear and bending-moment diagrams for the beam and loading shown, and determine the location and magnitude of the maximum bending moment.

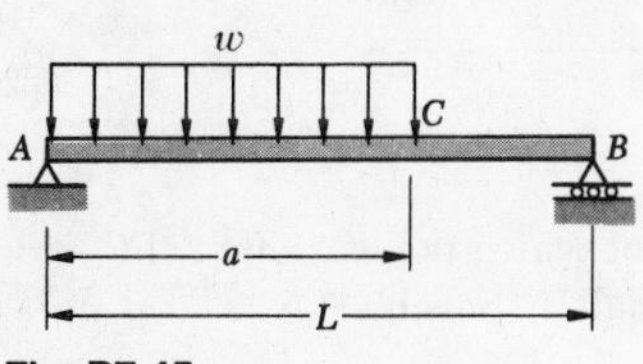

Fig. P7.45

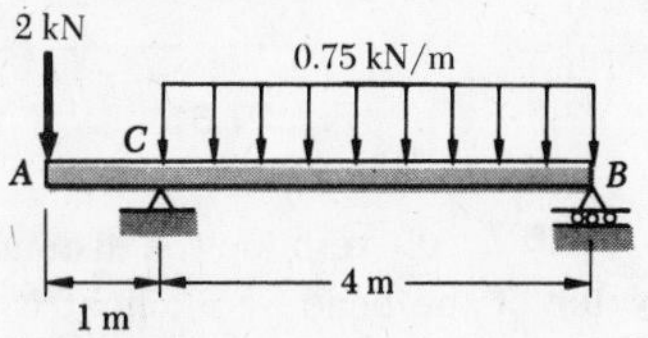

Fig. P7.46

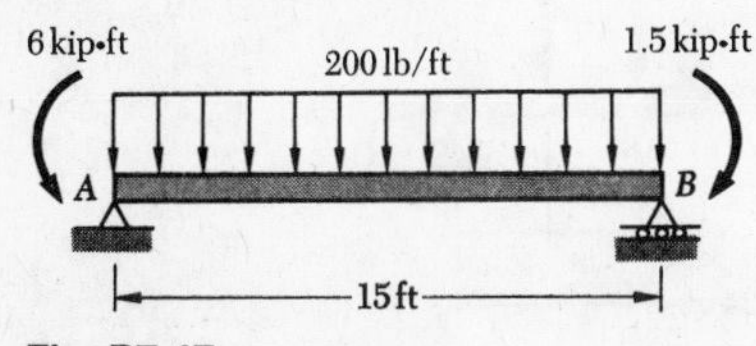

Fig. P7.47

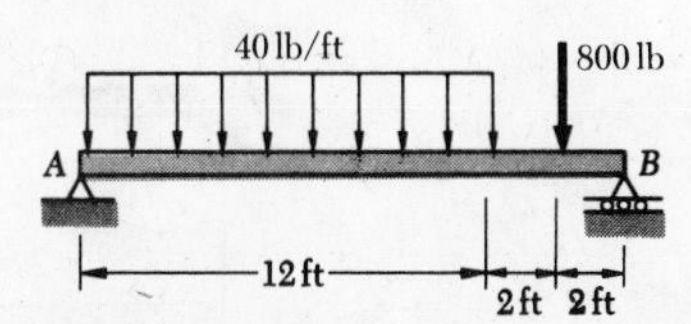

Fig. P7.48

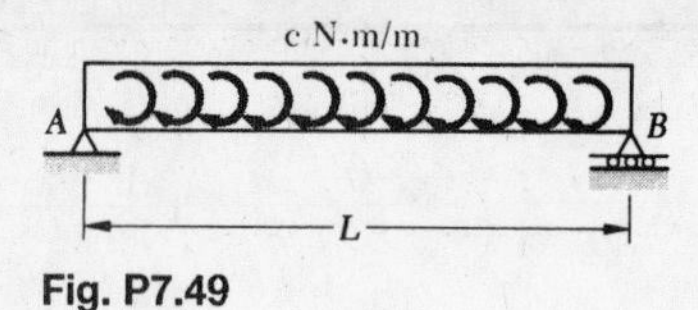

Fig. P7.49

7.49 A beam AB is loaded by couples spaced uniformly along its length. Assuming that the couples may be represented by a uniformly distributed couple loading of c N·m/m, draw the shear and bending-moment diagrams for the beam when it is supported (*a*) as shown, (*b*) as a cantilever with a fixed support at A and no support at B.

***7.50** The beam AB is acted upon by the uniformly distributed load of 1.25 kN/m and by two forces **P** and **Q**. It has been experimentally determined that the bending moment is + 800 N·m at point D and + 300 N·m at point E. Draw the shear and bending-moment diagrams for the beam.

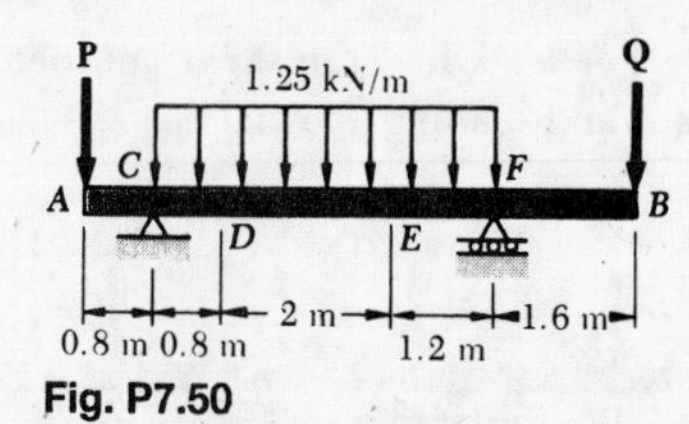

Fig. P7.50

***7.51** Concrete piles are designed primarily to resist axial loads, and their bending resistance is relatively small. To lessen the possibility of breaking them, piles are often lifted at several points along their length. By using the arrangement shown, the concrete pile AB is lifted by four equal forces applied at points C, D, E, and F. If the total weight of the pile is 20,000 lb, draw the shear and bending-moment diagrams of the pile. Assume the weight of the pile to be uniformly distributed.

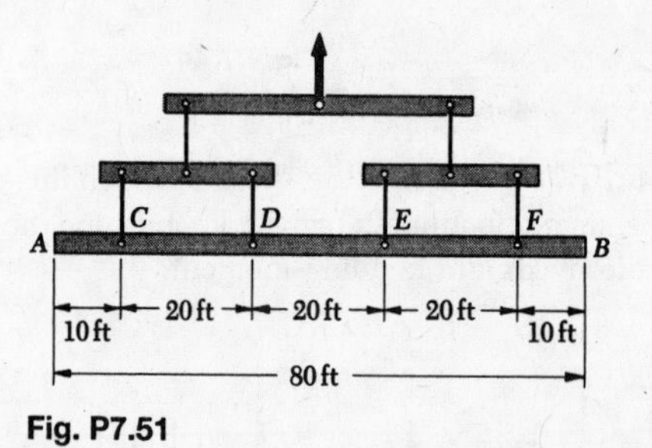

Fig. P7.51

7.52 Determine the distance a for which the maximum absolute value of the bending moment is as small as possible.

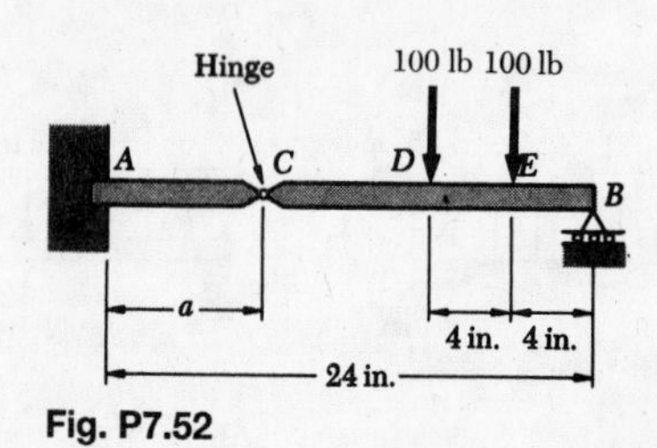

Fig. P7.52

SECTIONS 7.7 to 7.9

7.53 Three loads are suspended as shown from the cable. Knowing that $h_C = 12$ ft, determine (*a*) the components of the reaction at *E*, (*b*) the maximum value of the tension in the cable.

7.54 Determine the sag at point *C* if the maximum tension in the cable is 2600 lb.

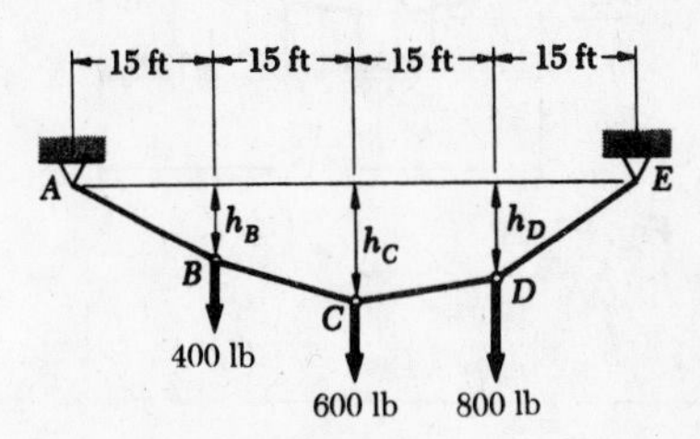

Fig. P7.53 and P7.54

7.55 If $a = 2$ m and $b = 2.25$ m, determine the components of the reaction at *E* for the cable and loading shown.

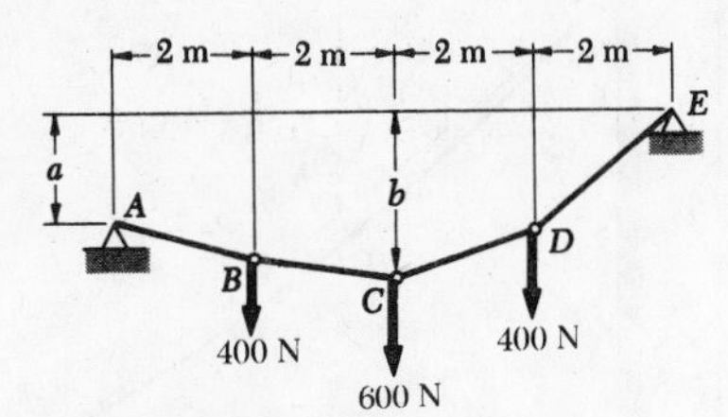

Fig. P7.55 and P7.56

7.56 Determine the distance *a* if the portion *BC* of the cable is horizontal and if the maximum tension in the cable is 2600 N.

7.57 An electric wire, weighing 0.20 lb/ft, is strung between two insulators at the same elevation and 100 ft apart. If the maximum tension in the wire is to be 80 lb, determine the smallest value of the sag which may be used. (Assume the wire to be parabolic.)

7.58 The central portion of cable *AB* is supported by a frictionless horizontal surface. Knowing that the mass per unit length of the cable is 3 kg/m and assuming the cable to be parabolic, determine the magnitude of the load **P** when $a = 8$ m.

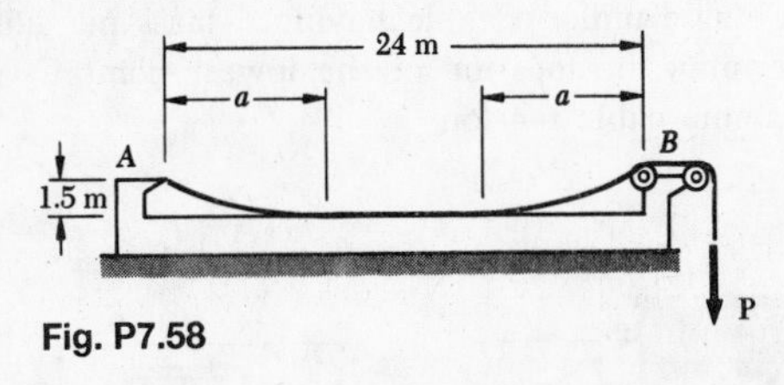

Fig. P7.58

7.59 Solve Prob. 7.58 when (*a*) $a = 6$ m, (*b*) $a = 12$ m.

7.60 Two cables of the same gage are attached to a transmission tower at *B*. Since the tower is slender, the horizontal component of the resultant of the forces exerted by the cables at *B* is to be zero. Assuming the cables to be parabolic, determine the required sag *h* of cable *AB*.

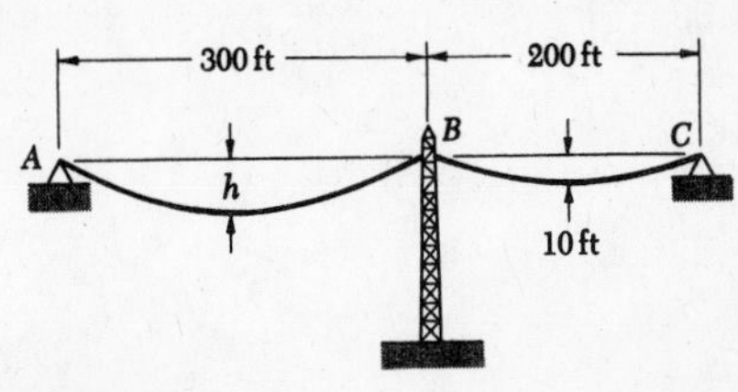

Fig. P7.60

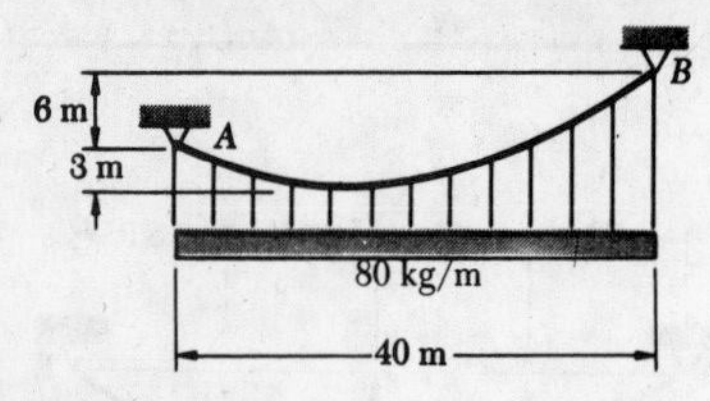

Fig. P7.61

7.61 Cable AB supports a load distributed uniformly along the horizontal as shown. The lowest point of the cable is 3 m below the support A. Determine the maximum and minimum values of the tension in the cable.

***7.62** The total weight of cable AC is 60 lb. Assuming that the weight of the cable is distributed uniformly along the horizontal, determine the sag h and the slope of the cable at A and C.

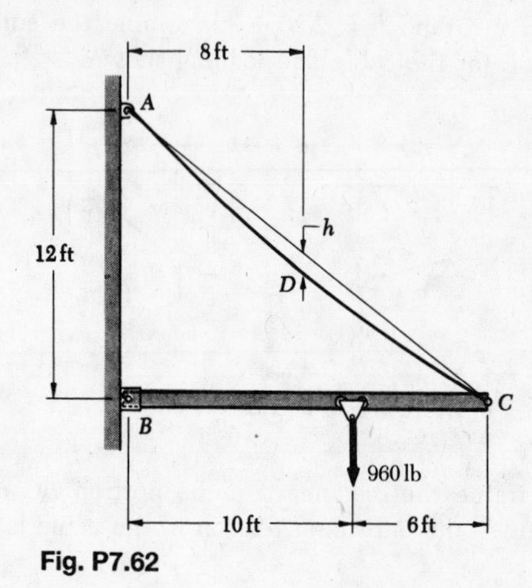

Fig. P7.62

***7.63** Solve Prob. 7.62, assuming that the total weight of cable AC is 120 lb.

7.64 A steam pipe has a mass per unit length of 70 kg/m. It passes between two buildings 20 m apart and is supported by a system of cables as shown. Assuming that the cable system causes the same loading as a single uniform cable having a mass per unit length of 5 kg/m, determine the location of the lowest point C of the cable and the maximum cable tension.

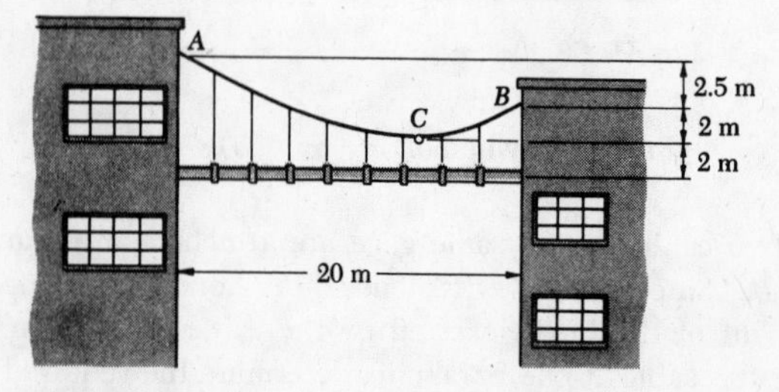

Fig. P7.64

SECTION 7.10

7.65 An aerial tramway cable of length 200 m and mass 1000 kg is suspended between two points at the same elevation. Knowing that the sag is 50 m, find the horizontal distance between supports and the maximum tension.

7.66 A 30-m rope is strung between the roofs of two buildings, each 9 m high. The maximum tension is found to be 250 N and the lowest point of the cable is observed to be 3 m above the ground. Determine the horizontal distance between the buildings and the total weight of the rope.

7.67 A $\frac{3}{4}$-in.-diameter wire rope weighing 0.90 lb/ft is suspended from two supports at the same elevation and 300 ft apart. If the sag is 75 ft, determine (*a*) the total length of the cable, (*b*) the maximum tension.

7.68 A copper transmission wire weighs 0.50 lb/ft and is attached to two insulators at the same elevation and 300 ft apart. It has been established that the horizontal component of the tension in the wire must be 175 lb if the insulators are not to be bent. Determine (*a*) the length of cable which should be used, (*b*) the resulting sag, (*c*) the resulting maximum tension.

7.69 A chain of length 10 m and total mass 20 kg is suspended between two points at the same elevation and 5 m apart. Determine the sag and the maximum tension.

7.70 The 4-m cable *AB* has a total mass of 10 kg and is attached to collars at *A* and *B* which may slide freely on the rods shown. Neglecting the weight of the collars, determine (*a*) the magnitude of the horizontal force **F** so that $h = a$, (*b*) the corresponding value of h and a, (*c*) the maximum tension in the cable.

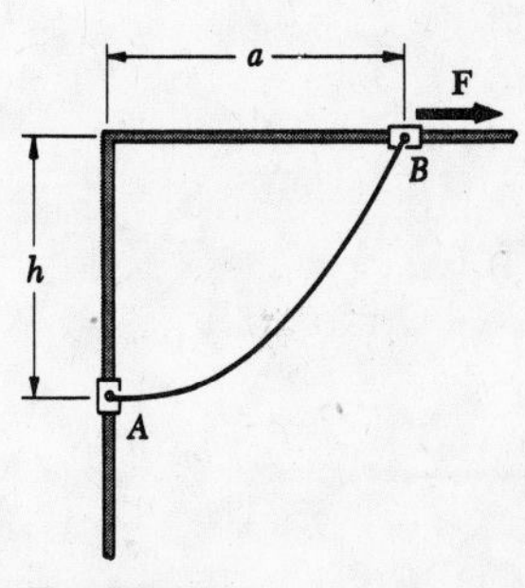

Fig. P7.70

7.71 A chain of length 20 ft and total weight 40 lb is suspended between two points at the same elevation and 10 ft apart. Determine the sag and the maximum tension.

***7.72** A cable, of weight w per unit length, is looped over a cylinder and is in contact with the cylinder above points *A* and *B*. Knowing that $\theta = 30°$, determine (*a*) the length of the cable, (*b*) the tension in the cable at *C*. (*Hint.* Use the property indicated in Prob. 7.112, page 311 of the text.)

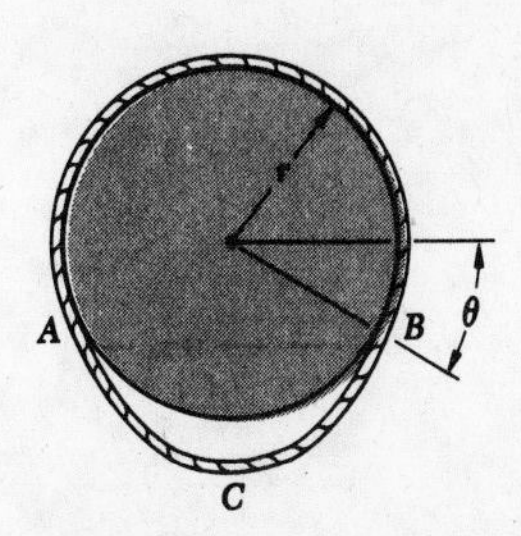

Fig. P7.72

CHAPTER 8
FRICTION

SECTIONS 8.1 to 8.4

8.1 A support block is acted upon by the two forces shown. Determine the magnitude of **P** required to start the block up the plane.

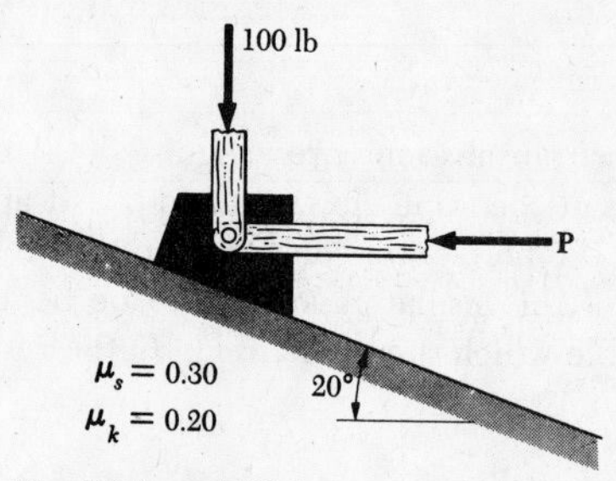

Fig. P8.1 and P8.2

8.2 Determine the smallest magnitude of the force **P** which will prevent the support block from sliding down the plane.

8.3 Denoting by ϕ_s the angle of static friction between the block and the plane, determine the magnitude and direction of the smallest force **P** which will cause the block to move up the plane.

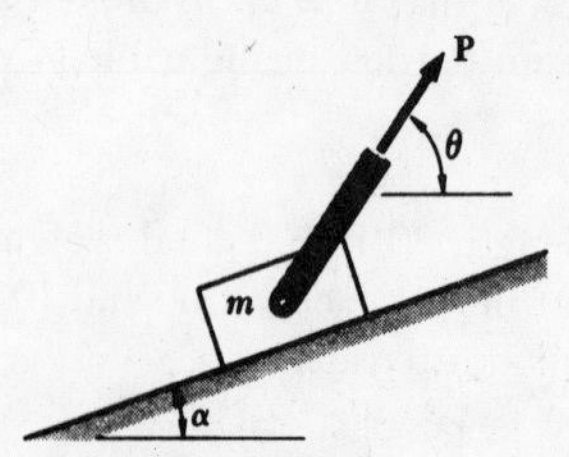

Fig. P8.3 and P8.4

8.4 A block of mass $m = 20$ kg rests on a rough plane as shown. Knowing that $\alpha = 25°$ and $\mu_s = 0.20$, determine the magnitude and direction of the smallest force **P** required (*a*) to start the block up the plane, (*b*) to prevent the block from moving down the plane.

8.5 The two 25-lb boxes A and B may be attached to the lever by means of either one or two horizontal links. The coefficient of friction between all surfaces is 0.30. Determine the magnitude of the force **P** required to move the lever (*a*) if a single link EF is used, (*b*) if a single link CD is used, (*c*) if two links EF and CD are used at the same time.

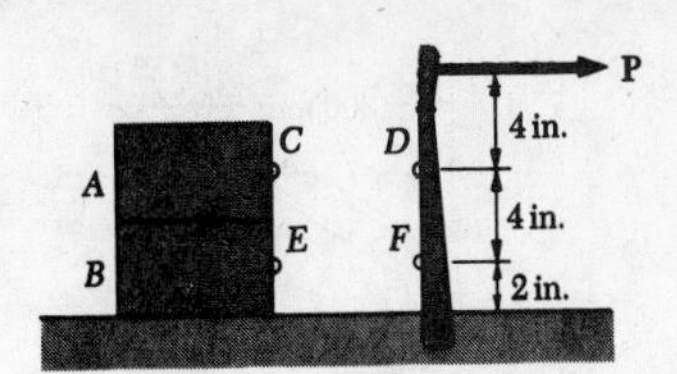

Fig. P8.5

8.6 Two packages are placed on a conveyor belt which is at rest. Between the belt and package A the coefficients of friction are $\mu_s = 0.2$ and $\mu_k = 0.15$; between package B and the belt the coefficients are $\mu_s = 0.3$ and $\mu_k = 0.25$. The packages are placed on the belt so that they are in contact with each other and at rest. Determine (*a*) whether either, or both, of the packages will move, (*b*) the friction force acting on each package.

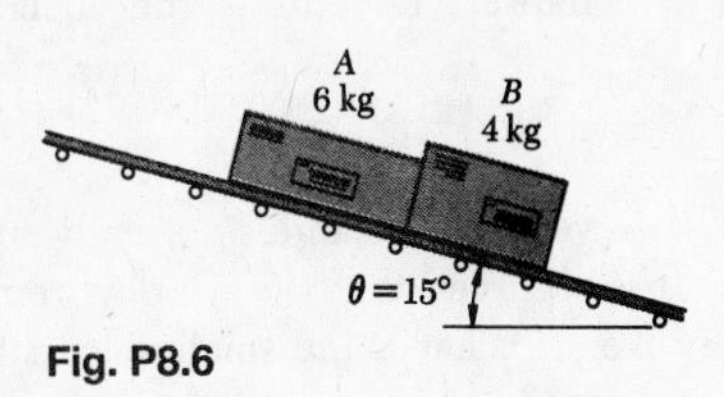

Fig. P8.6

8.7 Knowing that $\mu = 0.20$ at all surfaces of contact, determine the magnitude of the force **P** required to move the 10-kg plate B to the left. (Neglect bearing friction in the pulley.)

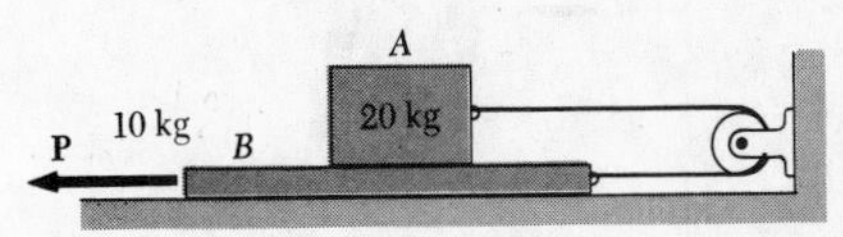

Fig. P8.7

8.8 Two blocks are connected by a cable as shown. Knowing that the coefficient of static friction between block A and the horizontal surface is 0.50, determine the range of values of θ for which the blocks will not move.

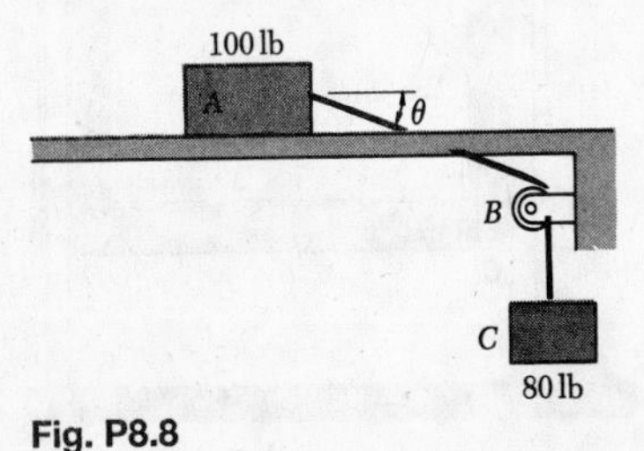

Fig. P8.8

8.9 A 180-lb sliding door is mounted on a horizontal rail as shown. The coefficients of static friction between the rail and the door at A and B are 0.20 and 0.30, respectively. Determine the horizontal force which must be applied to the handle C in order to move the door to the left.

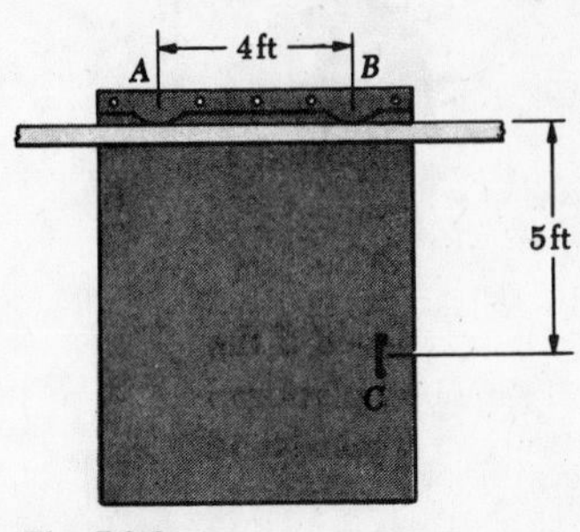

Fig. P8.9

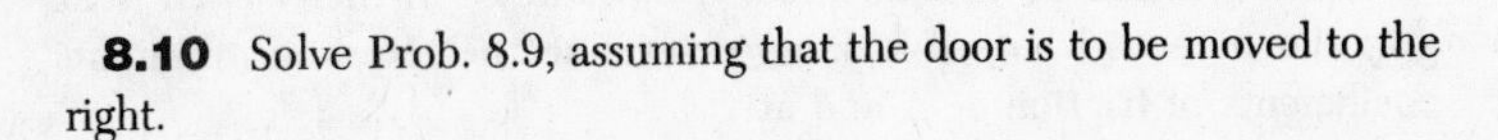

8.10 Solve Prob. 8.9, assuming that the door is to be moved to the right.

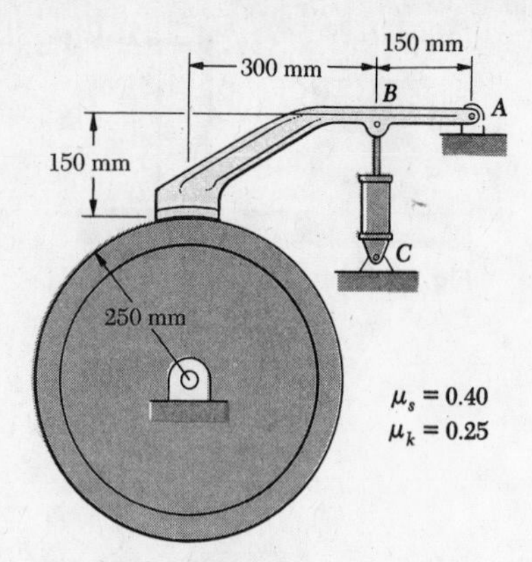

Fig. P8.11 and P8.12

8.11 A couple of magnitude 90 N · m is applied to the drum. Determine the smallest force which must be exerted by the hydraulic cylinder if the drum is not to rotate, when the applied couple is directed (*a*) clockwise, (*b*) counterclockwise.

8.12 The hydraulic cylinder exerts on point *B* a force of 2400 N directed downward. Determine the moment of the friction force about the axle of the drum when the drum is rotating (*a*) clockwise, (*b*) counterclockwise.

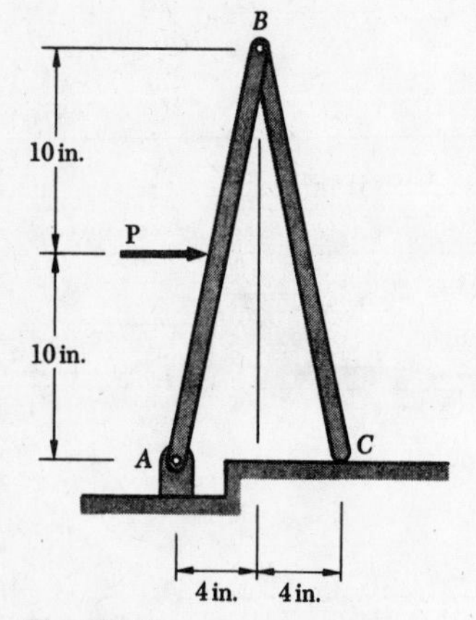

Fig. P8.13

8.13 Two uniform rods, each of weight 10 lb, are held by frictionless pins *A* and *B*. It is observed that if the magnitude of **P** is larger than 24 lb, the rods will collapse. Determine the coefficient of friction at *C*.

8.14 A window sash weighs 10 lb and is normally supported by two 5-lb sash weights. It is observed that the window remains open after one sash cord has broken. What is the smallest possible value of the coefficient of static friction? (Assume that the sash is slightly smaller than the frame and will bind only at points *A* and *D*.)

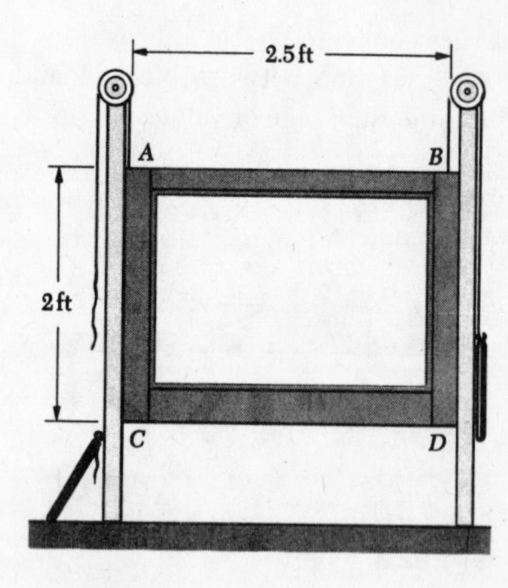

Fig. P8.14

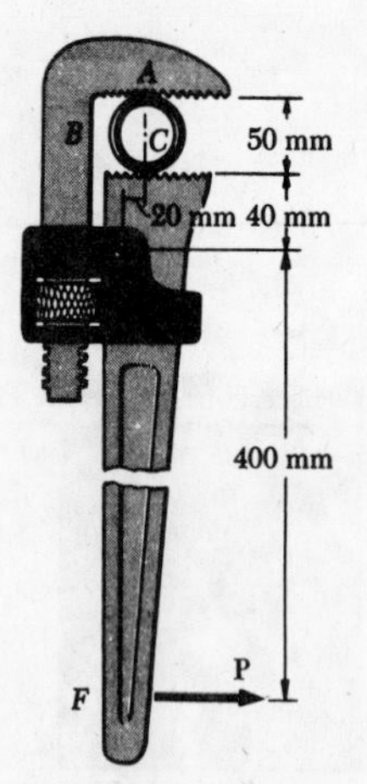

Fig. P8.15

8.15 A pipe of diameter 50 mm is gripped by the stillson wrench shown. Portions *AB* and *DE* of the wrench are rigidly attached to each other and portion *CF* is connected by a pin at *D*. If the wrench is to grip the pipe and be self-locking, determine the required minimum coefficients of friction at *A* and at *C*.

8.16 Two uniform rods, each of weight W, are held by frictionless pins A and B. It is observed that if the value of θ is greater than 10°, the rods will not remain in equilibrium. Determine the coefficient of friction at C.

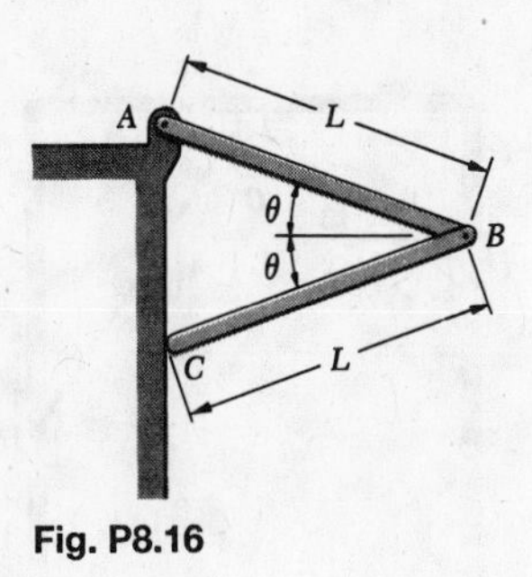

Fig. P8.16

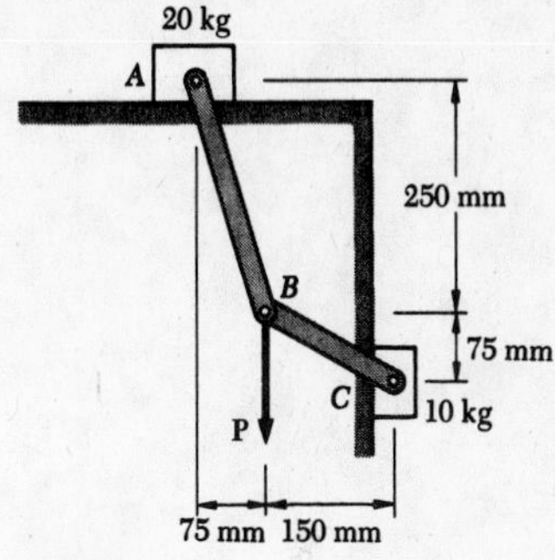

Fig. P8.17

8.17 Two links, of negligible weight, are connected by frictionless pins to each other and to two blocks at A and C. The coefficient of friction is 0.25 at A and C. If neither block is to slip, determine the magnitude of the largest force **P** which can be applied at B.

8.18 In Prob. 8.17, determine the magnitude of the smallest force **P** which must be applied at B if neither block is to slip.

8.19 The 10-lb uniform rod AB is held in the position shown by the force **P.** Knowing that the coefficient of friction is 0.20 at A and B, determine the smallest value of P for which equilibrium is maintained.

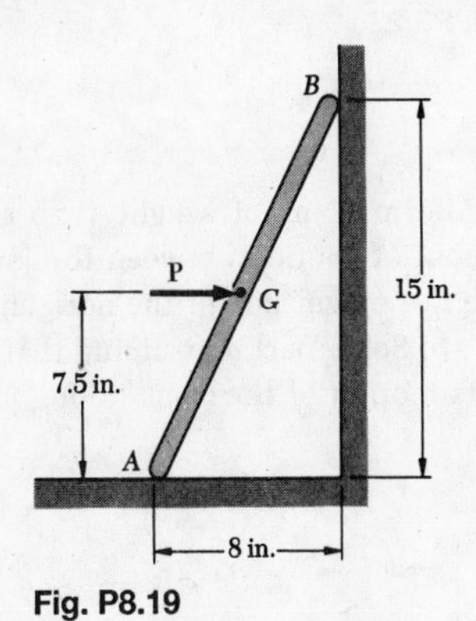

Fig. P8.19

8.20 In Prob. 8.19, determine the largest value of **P** for which equilibrium is maintained.

8.21 Two collars, each of weight W, are connected by a cord which passes over a frictionless pulley at C. Determine the smallest value of μ between the collars and the vertical rods for which the system will remain in equilibrium in the position shown.

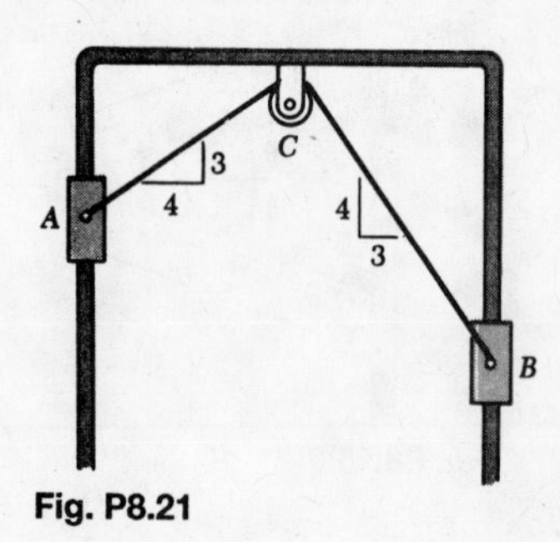

Fig. P8.21

8.22 The rod AB rests on a horizontal surface at A and against a sloping surface at B. Knowing that the coefficient of friction is 0.25 at both A and B, determine the maximum distance a at which the load **W** can be supported. Neglect the weight of the rod.

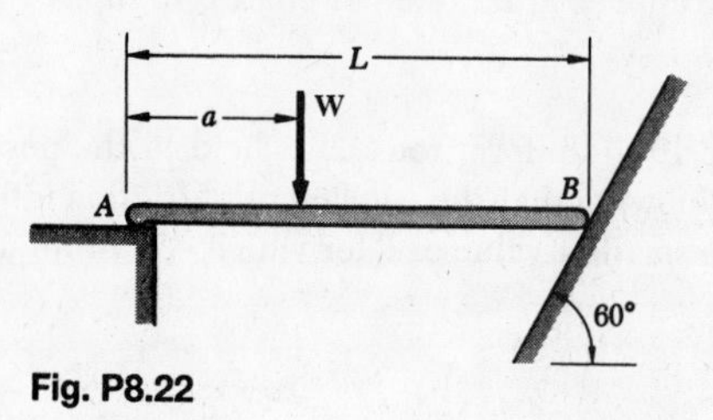

Fig. P8.22

8.23 A 10-ft uniform plank of weight 45 lb rests on two joists as shown. The coefficient of friction between the joists and the plank is 0.40. (*a*) Determine the magnitude of the horizontal force **P** required to move the plank. (*b*) Solve part *a* assuming that a single nail driven into joist A prevents motion of the plank along joist A.

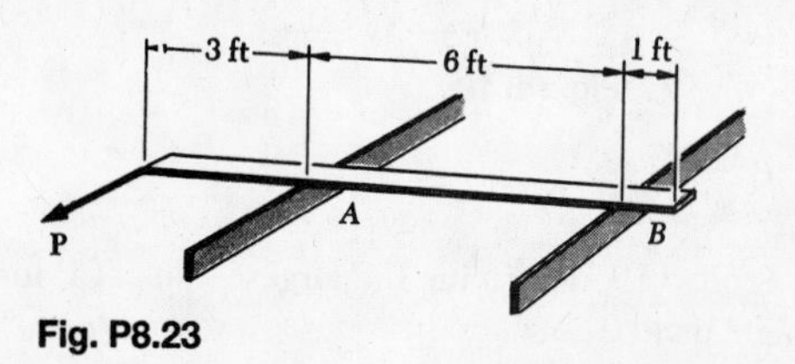

Fig. P8.23

8.24 Knowing that the coefficient of friction between the joists and the plank AB is 0.30, determine the smallest distance a for which the plank will slip at C.

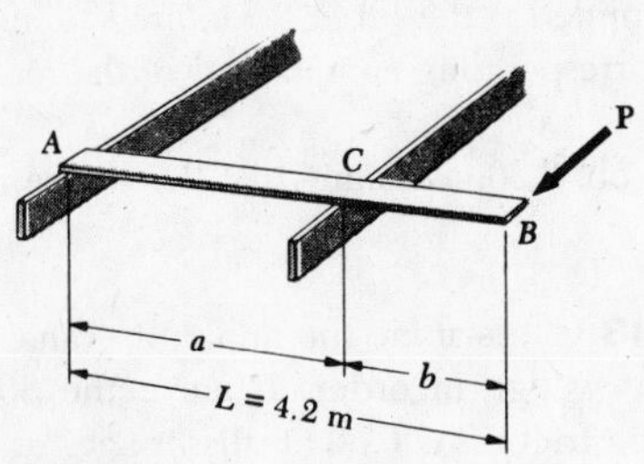

Fig. P8.24

8.25 A cylinder of weight W is placed in a V block as shown. Denoting by ϕ the angle of friction between the cylinder and the block and assuming $\phi < \theta$, determine (a) the axial force **P** required to move the cylinder, (b) the couple **M**, applied in the plane of the cross section of the cylinder, required to rotate the cylinder.

Fig. P8.25

8.26 Two rods are connected by a collar at B; a couple $\mathbf{M}_A$ of moment 250 lb·in. is applied to rod AB. Knowing that $\mu = 0.30$ between the collar and rod AB, determine the *maximum* couple $\mathbf{M}_C$ for which equilibrium will exist.

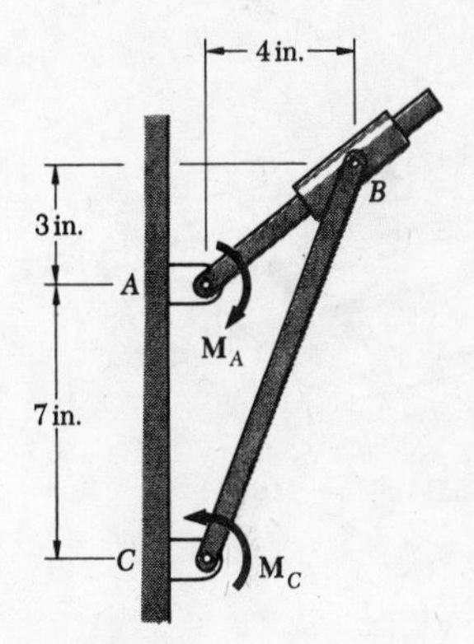

Fig. P8.26

8.27 In Prob. 8.26, determine the *minimum* couple $\mathbf{M}_C$ required for equilibrium.

8.28 Identical cylindrical cans, each of weight W, are raised to the top of an incline by a series of moving arms. Either one or two cans are moved by each arm. The coefficient of friction between all surfaces is $\mu = 0.20$. If $W = 2.00$ lb and $\theta = 12°$, determine the force parallel to the incline which the arm must exert on can A to move it. Does can A roll or slide?

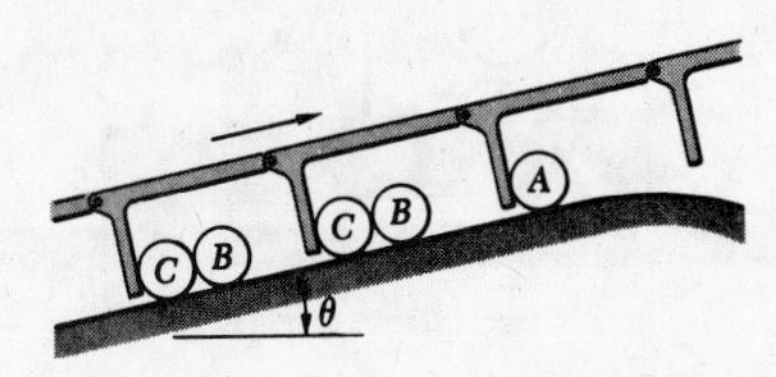

Fig. P8.28

***8.29** Solve Prob. 8.28, considering can C instead of can A.

SECTIONS 8.5 and 8.6

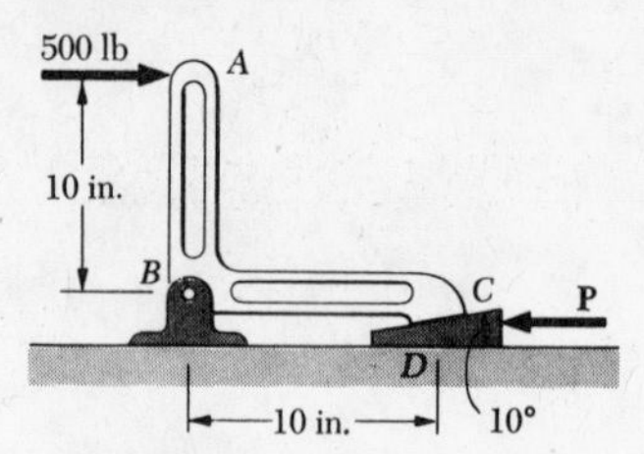

Fig. P8.30

8.30 The machine part ABC is supported by a frictionless hinge at B and by a wedge at C. Knowing that $\mu = 0.20$ at both surfaces of the wedge, determine (*a*) the force **P** required to move the wedge to the left, (*b*) the corresponding components of the reaction at B.

8.31 Solve Prob. 8.30, assuming that the wedge is to be moved to the right.

8.32 and 8.33 Determine the minimum value of **P** which must be applied to the wedge in order to move the 800-kg block. The coefficient of static friction is 0.30 at all surfaces of contact.

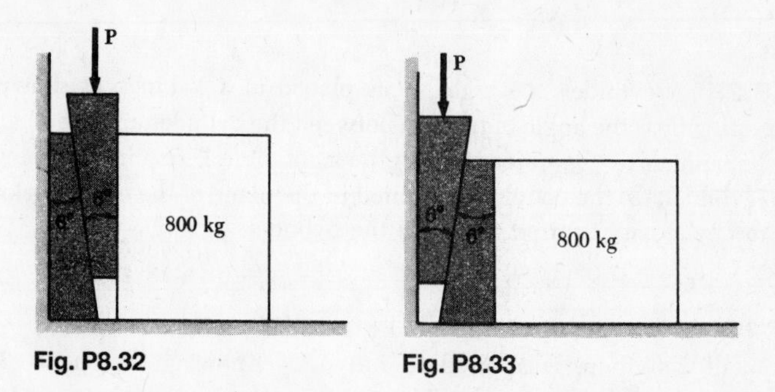

Fig. P8.32 Fig. P8.33

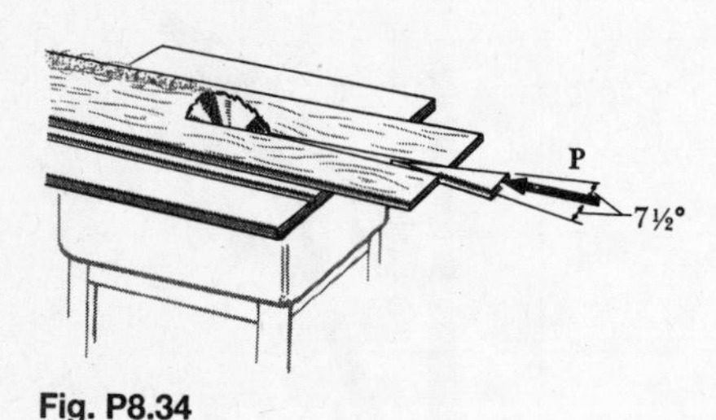

Fig. P8.34

8.34 A 15° wedge is forced into a saw cut to prevent binding of the circular saw. The coefficient of friction between the wedge and the wood is 0.25. Knowing that a horizontal force **P** of magnitude 150 N was required to insert the wedge, determine the magnitude of the forces exerted on the board by the wedge after it has been inserted.

8.35 A 5° wedge is to be forced under a machine base at A. Knowing that $\mu = 0.15$ at all surfaces, (*a*) determine the force **P** required to move the wedge, (*b*) indicate whether the machine will move.

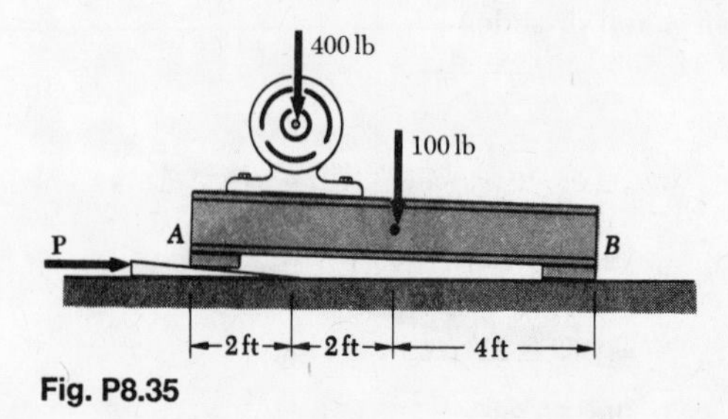

Fig. P8.35

8.36 Solve Prob. 8.35, assuming that the wedge is to be forced under the machine base at B instead of A.

8.37 A conical wedge is placed between two horizontal plates which are then slowly moved toward each other. Indicate what will happen to the wedge (*a*) if $\mu = 0.15$, (*b*) if $\mu = 0.25$.

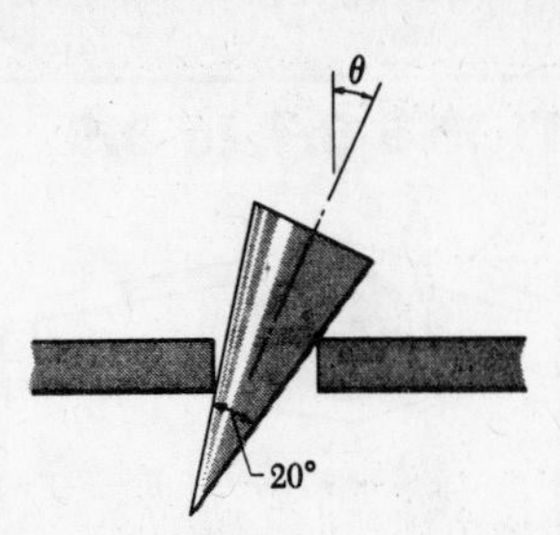

Fig. P8.37

8.38 The square-threaded worm gear shown has a mean radius of 40 mm and a pitch of 10 mm. The large gear is subjected to a constant clockwise torque of 900 N · m. Knowing that the coefficient of friction between gear teeth is 0.10, determine the torque which must be applied to shaft *AB* in order to rotate the large gear counterclockwise. Neglect friction in the bearings at *A*, *B*, and *C*.

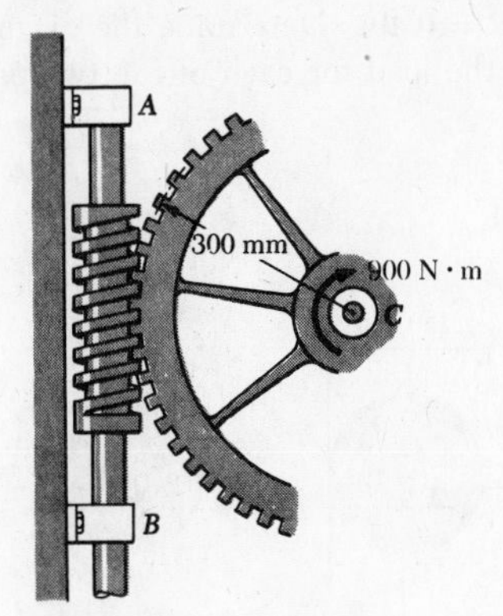

Fig. P8.38

8.39 In Prob. 8.38, determine the torque which must be applied to shaft *AB* in order to rotate the large gear clockwise.

8.40 The ends of two fixed rods *A* and *B* are each made in the form of a single-threaded screw of mean radius 0.28 in. and pitch 0.09 in. The coefficient of friction between the rods and the threaded sleeve is 0.15. Determine the moment of the couple which must be applied to the sleeve in order to draw the rods closer together. Rod *A* has a left-handed thread and rod *B* a right-handed thread.

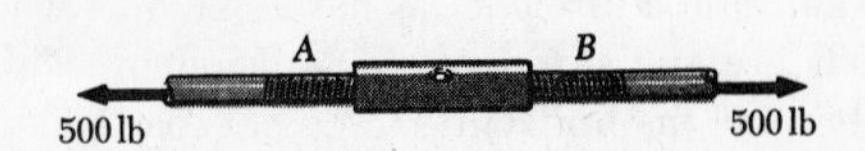

Fig. P8.40

8.41 In Prob. 8.40, a right-handed thread is used on *both* rods *A* and *B*. Determine the moment of the couple which must be applied to the sleeve in order to rotate it.

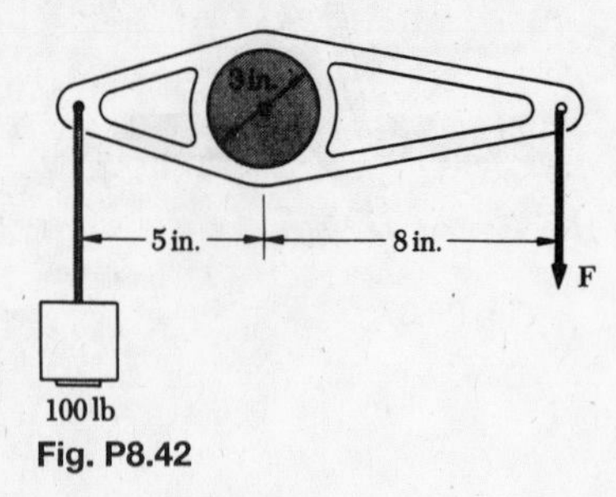

Fig. P8.42

8.42 A lever of negligible weight is loosely fitted onto a 3-in.-diameter fixed shaft as shown. It is observed that a force **F** of magnitude 70 lb will just start rotating the lever clockwise. Determine (*a*) the coefficient of friction between the shaft and the lever, (*b*) the smallest force **F** for which the lever does not start rotating counterclockwise.

8.43 A windlass, of diameter 150 mm, is used to raise a 40-kg load. It is supported by two axles and bearings of diameter 50 mm, poorly lubricated ($\mu = 0.40$). Determine the magnitude of the force **P** required to raise the load for each of the two positions shown.

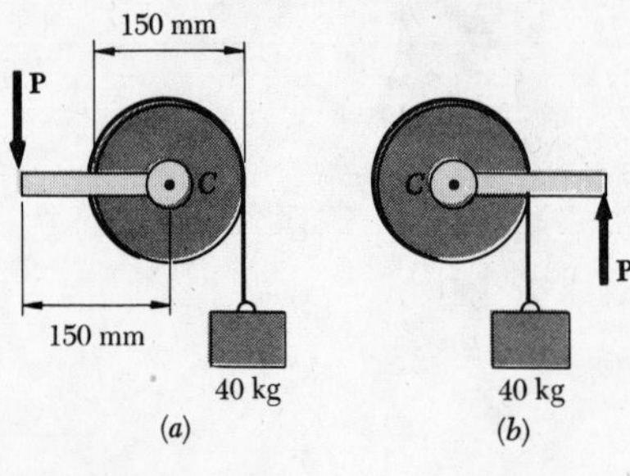

Fig. P8.43

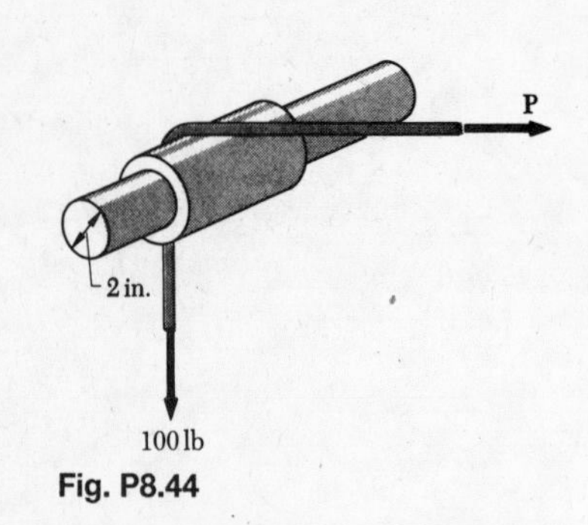

Fig. P8.44

8.44 A bushing of outside diameter 3 in. fits loosely on a horizontal 2-in.-diameter shaft. A horizontal force **P** of magnitude 110 lb is required to raise the 100-lb load attached to the rope. Determine the coefficient of friction between the shaft and the bushing. Assume that the rope does not slip on the bushing.

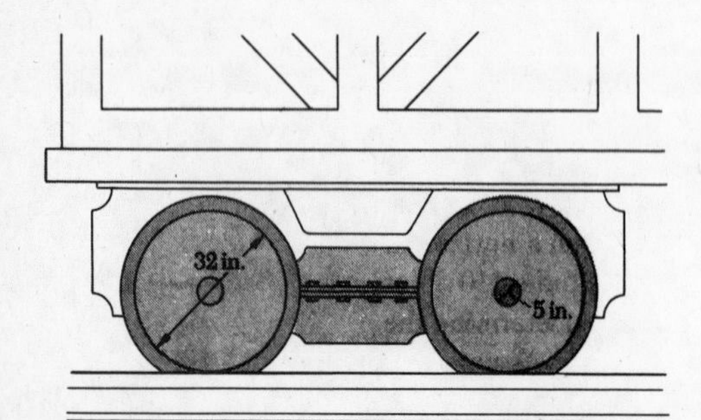

Fig. P8.45

8.45 A certain railroad freight car has eight steel wheels of 32-in. diameter which are supported on 5-in.-diameter axles. Assuming $\mu = 0.015$, determine the horizontal force per ton of load required to move the car at constant velocity.

8.46 A scooter is to be designed to roll down a 2 percent slope at a constant speed. Assuming that the coefficient of kinetic friction between the 24-mm axles and the bearings is 0.10, determine the required diameter of the wheels. Neglect the rolling resistance between the wheels and the ground.

8.47 A couple of magnitude 150 lb · ft is required to start the vertical shaft rotating. Determine the coefficient of static friction.

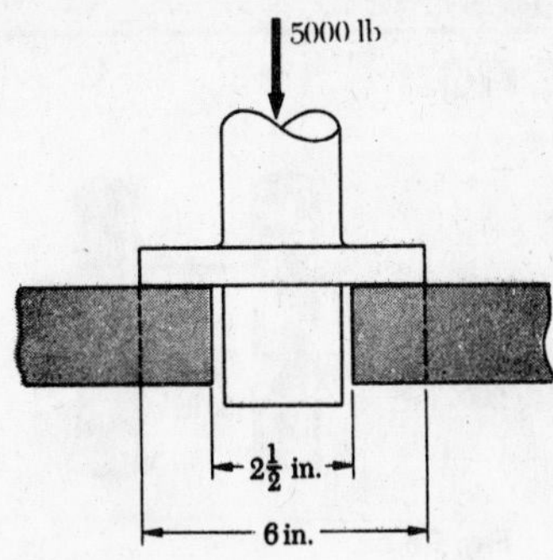

Fig. P8.47

8.48 A 50-lb electric floor polisher is operated on a surface for which the coefficient of friction is 0.25. Assuming the normal force per unit area between the disk and the floor to be uniform, determine the magnitude Q of the horizontal forces required to prevent motion of the machine.

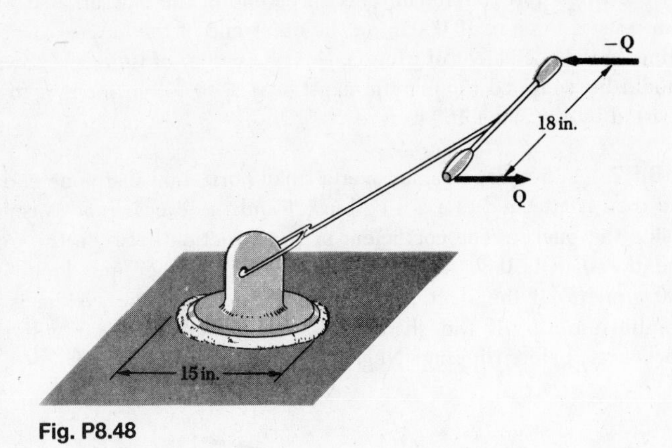

Fig. P8.48

8.49 In the clutch shown, disks A and B are keyed to the shaft but are free to slide along it. Disks C, D, and E are free to move parallel to the shaft but cannot rotate. The coefficient of static friction is 0.25 between all surfaces in contact. If disks C and E are pressed against the other disks as shown, determine the magnitude of the couple **M** required to rotate the shaft.

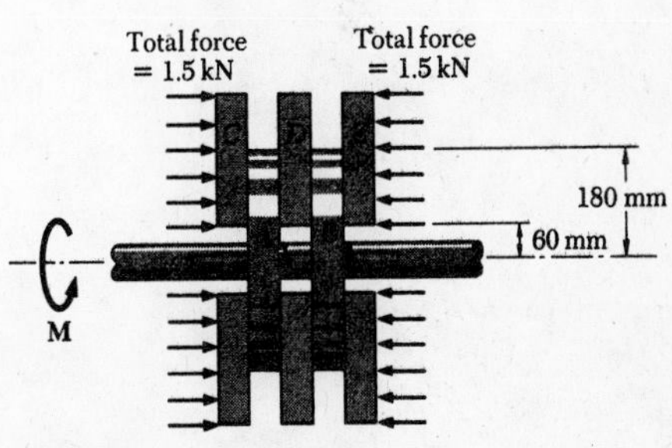

Fig. P8.49

8.50 In Prob. 8.49, determine the smallest forces with which disks C and E must be pressed against the other disks if the shaft is not to rotate when $M = 75$ N · m.

8.51 Determine the horizontal force required to move a 1200 kg automobile along a horizontal road at a constant velocity. Neglect all forms of friction except rolling resistance, and assume the coefficient of rolling resistance to be 1.5 mm. The diameter of each tire is 600 mm.

8.52 A circular disk of diameter 8 in. rolls at constant velocity down an incline which has a slope of $\frac{1}{4}$ in. per foot. Determine the coefficient of rolling resistance.

8.53 Solve Prob. 8.45, including the effect of a coefficient of rolling resistance of 0.02 in.

8.54 Solve Prob. 8.46, including the effect of a coefficient of rolling resistance of 1.8 mm.

SECTION 8.10

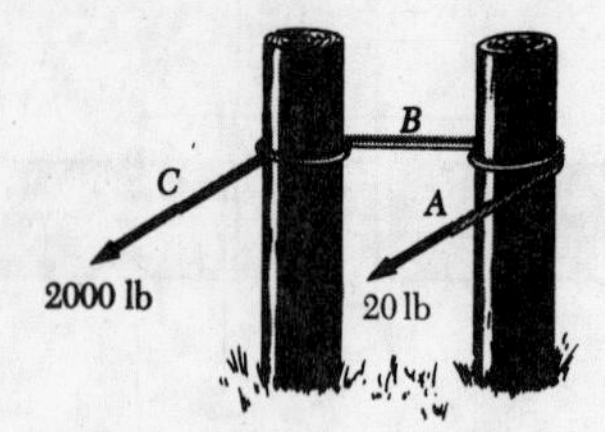

Fig. P8.55

8.55 A rope is wrapped around two posts as shown. If a 20-lb force must be exerted at A to resist a 2000-lb force at C, determine (*a*) the coefficient of static friction between the rope and the posts, (*b*) the corresponding tension in portion B of the rope.

8.56 A hawser is wrapped two full turns around a capstan head. By exerting a force of 160 lb on the free end of the hawser, a seaman can resist a force of 10,000 lb on the other end of the hawser. Determine (*a*) the coefficient of friction, (*b*) the number of times the hawser should be wrapped around the capstan if a 40,000-lb force is to be resisted by the same 160-lb force.

8.57 A 15-m rope passes over a small horizontal shaft; one end of the rope is attached to a 2.4-kg bucket and the excess rope is coiled inside the bucket. The coefficient of static friction between the rope and the shaft is 0.30, and the rope has a mass per unit length of 800 g/m. (*a*) If the shaft is held fixed, show that the system is in equilibrium. (*b*) If the shaft is slowly rotated, how far will the bucket rise before slipping? Neglect the diameter of the shaft.

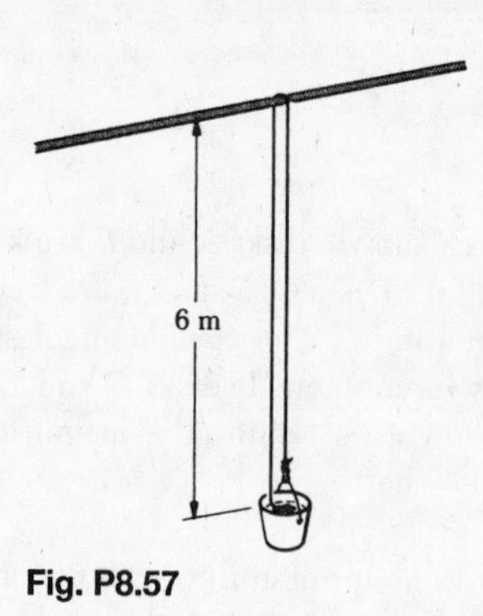

Fig. P8.57

8.58 The shaft of Prob. 8.57 is slowly rotated. Determine how far the bucket can be lowered before the rope slips on the shaft.

8.59 Assume that the bushing of Prob. 8.44 has become frozen to the shaft and cannot rotate. Determine the coefficient of friction between the bushing and the rope if a force **P** of magnitude 120 lb is required to raise the 100-lb load.

$\mu_s = 0.30$
$\mu_k = 0.25$

200 mm O 30° P C A B D

75 mm 175 mm

Fig. P8.60

8.60 A band brake is used to control the speed of a flywheel as shown. What torque must be applied to the flywheel in order to keep it rotating at a constant speed, when $P = 60$ N? Assume that the flywheel rotates clockwise.

8.61 Solve Prob. 8.60, assuming that the flywheel rotates counterclockwise.

8.62 A brake drum of radius $r = 150$ mm is rotating clockwise when a force **P** of magnitude 75 N is applied at D. Knowing that $\mu = 0.25$, determine the moment about O of the friction forces applied to the drum when $a = 75$ mm and $b = 400$ mm.

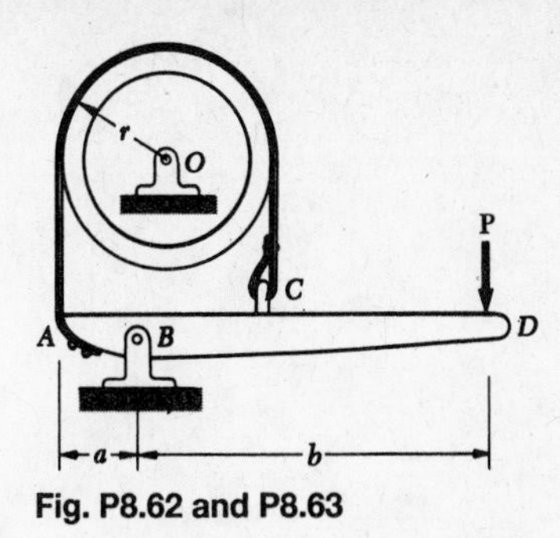

Fig. P8.62 and P8.63

8.63 Knowing that $r = 150$ mm and $a = 75$ mm, determine the maximum value of μ for which the brake is not self-locking. The brake drum rotates clockwise.

8.64 A flat belt is used to transmit a torque from pulley A to pulley B. The radius of each pulley is 2 in. and $\mu_s = 0.30$. Determine the largest torque which can be transmitted if the maximum allowable tension is 600 lb.

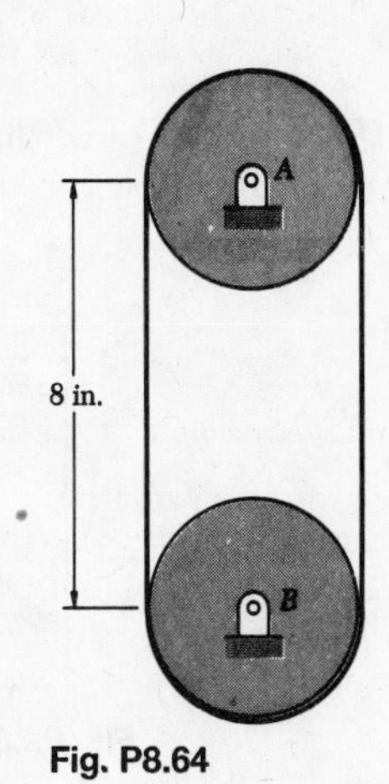

Fig. P8.64

8.65 Solve Prob. 8.64, assuming that the belt is looped around the pulleys in a figure 8.

8.66 A couple $\mathbf{M}_0$, of magnitude 150 lb · ft, must be applied to the drive pulley B in order to maintain a constant motion in the conveyor shown. Knowing that $\mu = 0.20$, determine the minimum tension in the lower portion of the belt if no slipping is to occur.

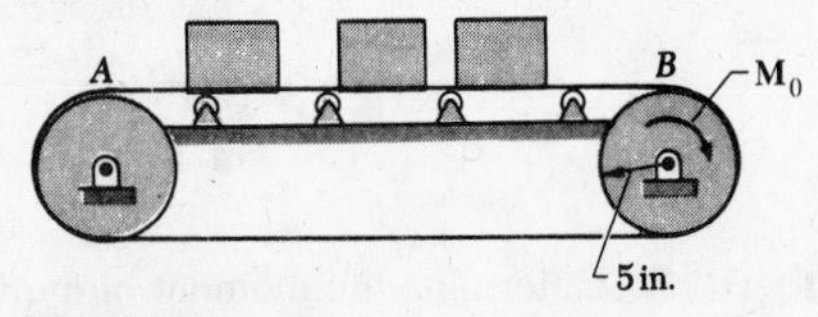

Fig. P8.66

CHAPTER 9
DISTRIBUTED FORCES: MOMENTS OF INERTIA

SECTIONS 9.1 to 9.5

9.1 through 9.3 Determine by direct integration the moment of inertia of the shaded area with respect to the y axis.

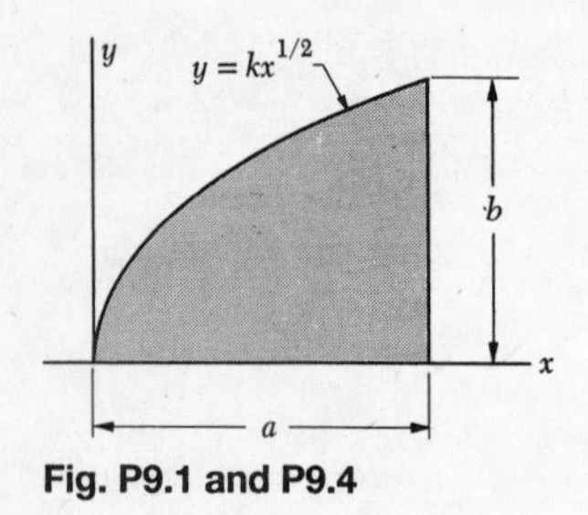

Fig. P9.1 and P9.4

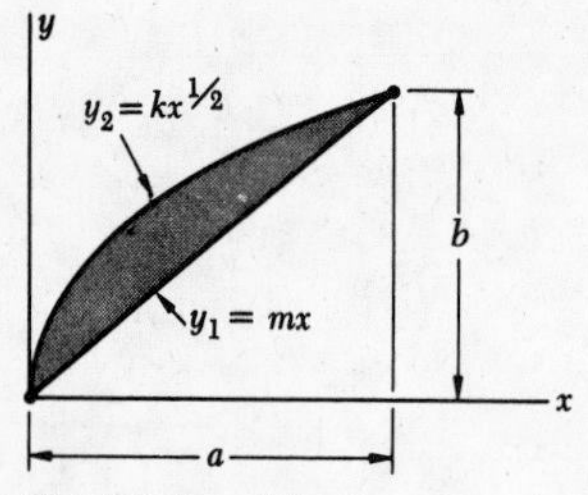

Fig. P9.2 and P9.5

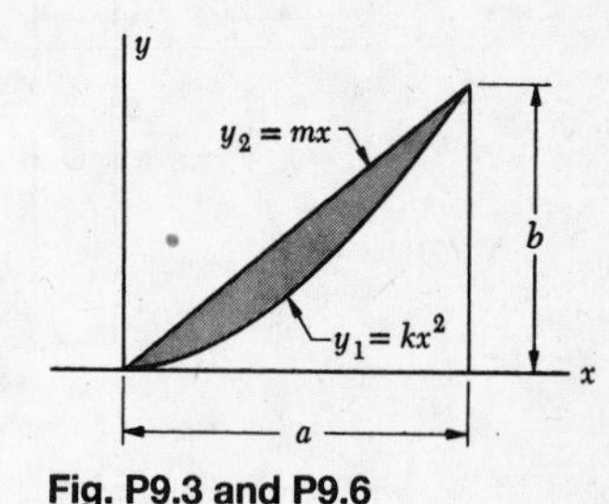

Fig. P9.3 and P9.6

9.4 through 9.6 Determine by direct integration the moment of inertia of the shaded area with respect to the x axis.

9.7 through 9.9 Determine the moment of inertia and radius of gyration of the shaded area shown with respect to the x axis.

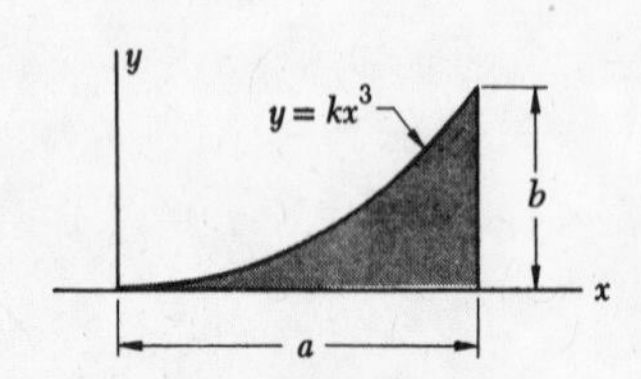

Fig. P9.7 and P9.10

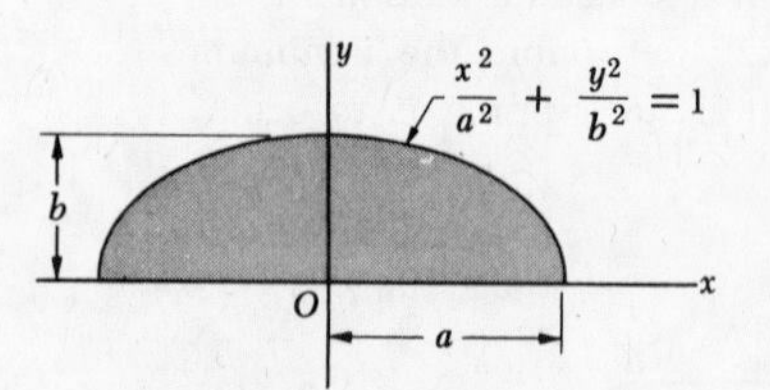

Fig. P9.8 and P9.11

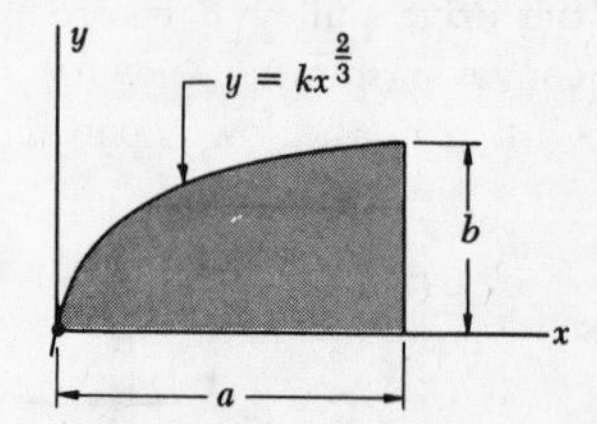

Fig. P9.9 and P9.12

9.10 through 9.12 Determine the moment of inertia and radius of gyration of the shaded area shown with respect to the y axis.

9.13 Determine the radius of gyration of an equilateral triangle of side a with respect to one of its sides.

9.14 Determine the polar moment of inertia and the polar radius of gyration of a rectangle of base b and height h with respect to one of its corners.

9.15 Determine the polar moment of inertia and polar radius of gyration of a square of side a with respect to the midpoint of one of its sides.

9.16 Determine the polar moment of inertia and the polar radius of gyration of the semielliptical area of Prob. 9.8 with respect to O.

***9.17** Determine the moment of inertia of the shaded area with respect to the x axis.

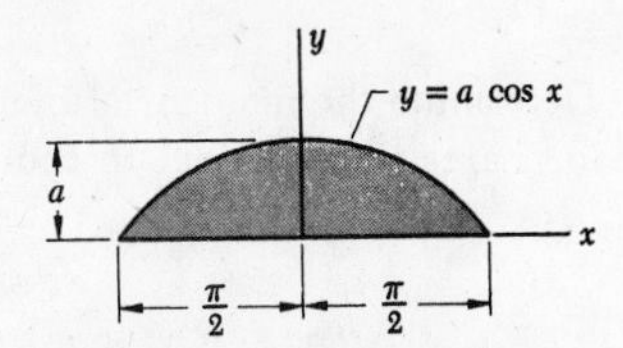

Fig. P9.17 and P9.18

***9.18** Determine the moment of inertia of the shaded area with respect to the y axis.

***9.19** In the plane area shown, it is desired to have I_x directly proportional to the height b. Determine the equation of the curve bounding the area on the right.

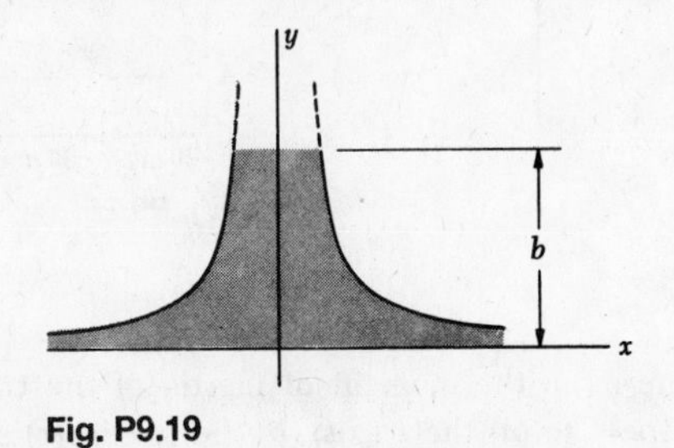

Fig. P9.19

SECTIONS 9.6 and 9.7

9.20 and 9.22 Determine the moment of inertia and the radius of gyration of the shaded area with respect to the x axis.

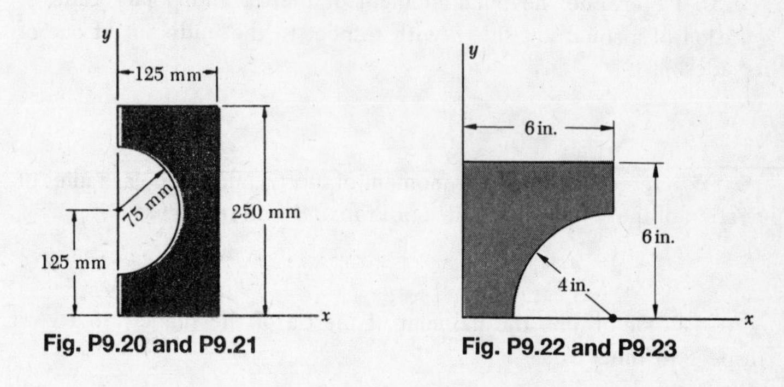

Fig. P9.20 and P9.21

Fig. P9.22 and P9.23

9.21 and 9.23 Determine the moment of inertia and the radius of gyration of the shaded area with respect to the y axis.

9.24 (*a*) Determine I_x and I_y if $b = 10$ in. (*b*) Determine the dimension b for which $I_x = I_y$.

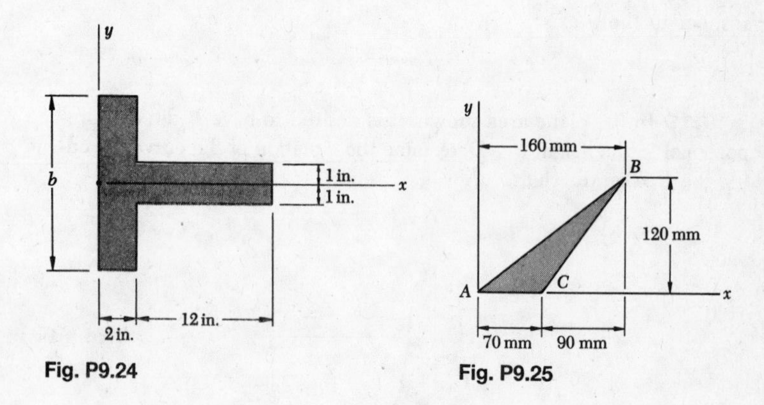

Fig. P9.24

Fig. P9.25

9.25 Determine the moment of inertia of the triangular area shown with respect to (*a*) the x axis, (*b*) the y axis, (*c*) side AB.

9.26 Determine the shaded area and its moment of inertia with respect to a centroidal axis parallel to AA', knowing that its moments of inertia with respect to AA' and BB' are respectively 800×10^6 mm^4 and 1.6×10^9 mm^4, and that $d_1 = 200$ mm and $d_2 = 100$ mm.

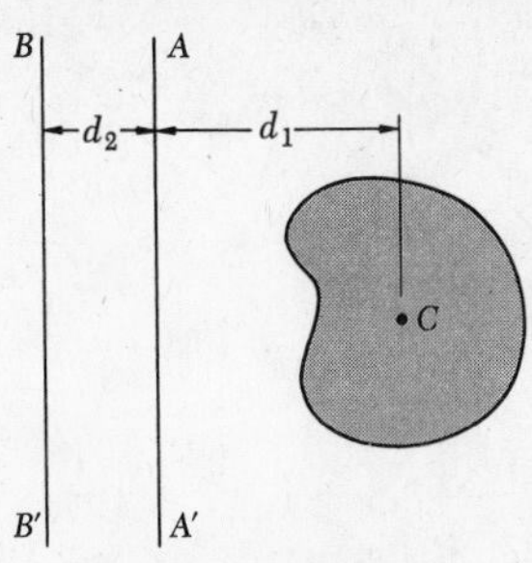

Fig. P9.26 and P9.27

9.27 Knowing that the shaded area is equal to 16×10^3 mm^2 and that its moment of inertia with respect to AA' is 325×10^6 mm^4, determine its moment of inertia with respect to BB', for $d_1 = 125$ mm and $d_2 = 50$ mm.

9.28 Determine the polar moment of inertia of the area shown with respect to (*a*) point O, (*b*) the centroid of the area.

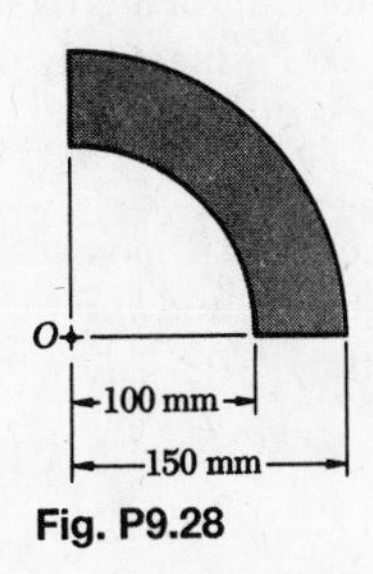

Fig. P9.28

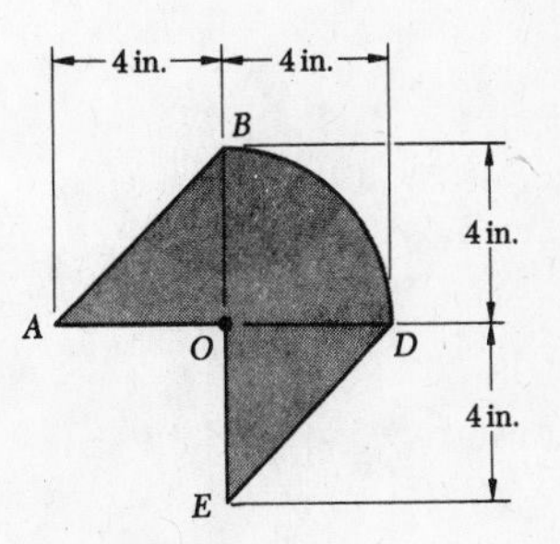

Fig. P9.29

9.29 Determine the polar moment of inertia of the area shown with respect to (*a*) point O, (*b*) the centroid of the area.

9.30 Determine the centroidal polar moment of inertia of the area shown.

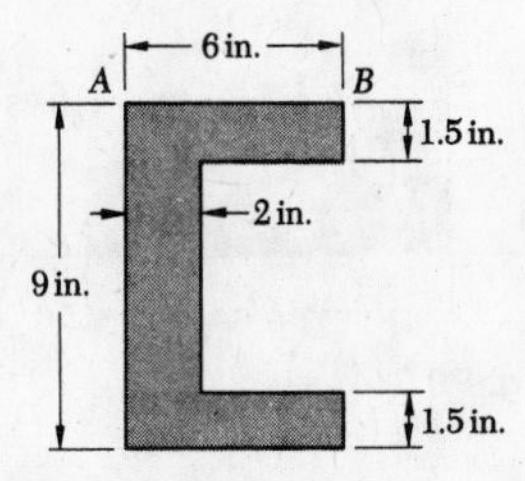

Fig. P9.30 and P9.31

9.31 Determine the moments of inertia $\bar{I}_x$ and $\bar{I}_y$ of the area shown with respect to centroidal axes respectively parallel and perpendicular to the side AB.

9.32 Determine the moments of inertia $\bar{I}_x$ and $\bar{I}_y$ of the area shown with respect to centroidal axes respectively parallel and perpendicular to the side AB.

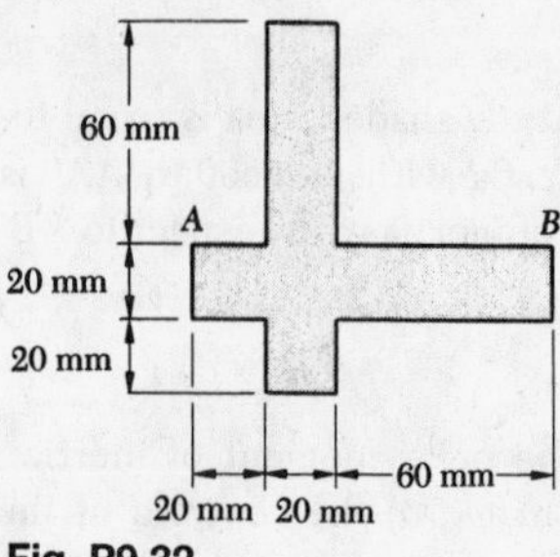

Fig. P9.32

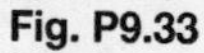
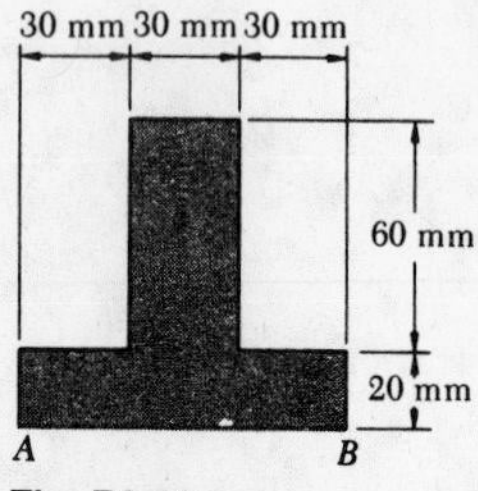

Fig. P9.33

9.33 Determine the moments of inertia $\bar{I}_x$ and $\bar{I}_y$ of the area shown with respect to centroidal axes respectively parallel and perpendicular to the side AB.

9.34 Determine the centroidal moments of inertia $\bar{I}_x$ and $\bar{I}_y$ and the centroidal radii of gyration $\bar{k}_x$ and $\bar{k}_y$ for the structural shape shown. Neglect the effect of fillets.

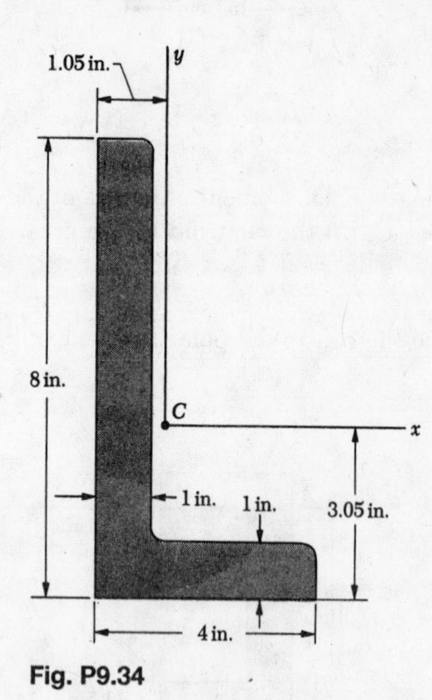

Fig. P9.34

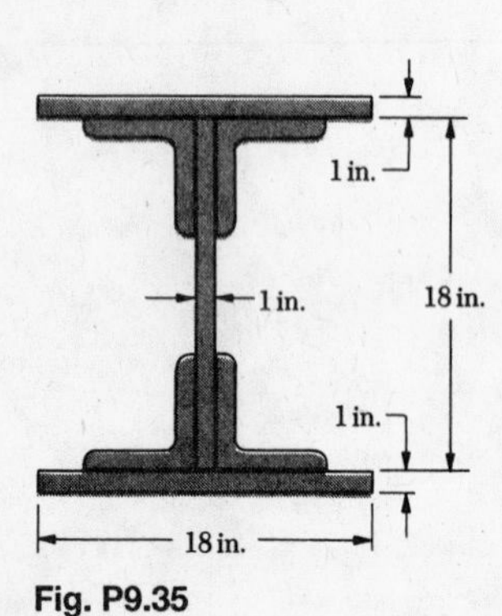

Fig. P9.35

9.35 Three steel plates, 18 by 1 in., are riveted to four 6- by 6-in. angles, each 1 in. thick, to form the column whose cross section is shown. Determine the moments of inertia and the radii of gyration of the section with respect to centroidal axes respectively parallel and perpendicular to the flanges.

SECTIONS 9.8 to 9.10

9.36 through 9.39 Using the parallel-axis theorem, determine the product of inertia of the area shown with respect to the centroidal x and y axes.

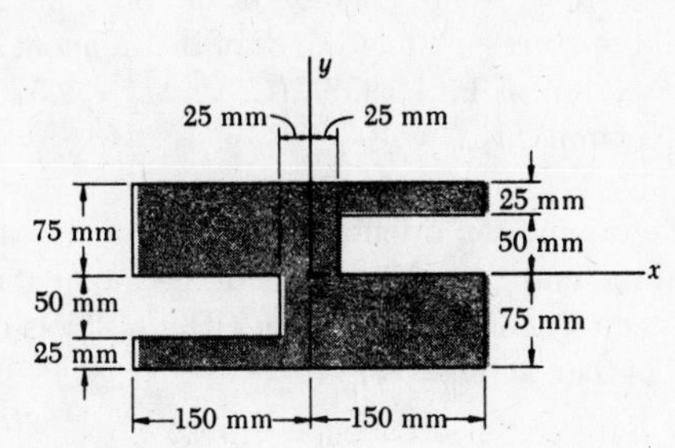

Fig. P9.36

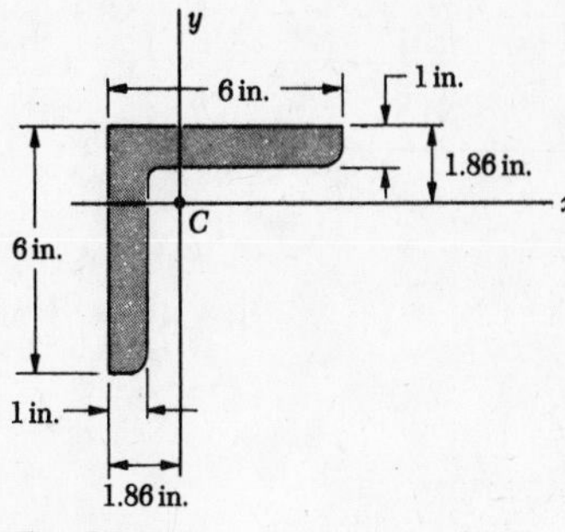

Fig. P9.37

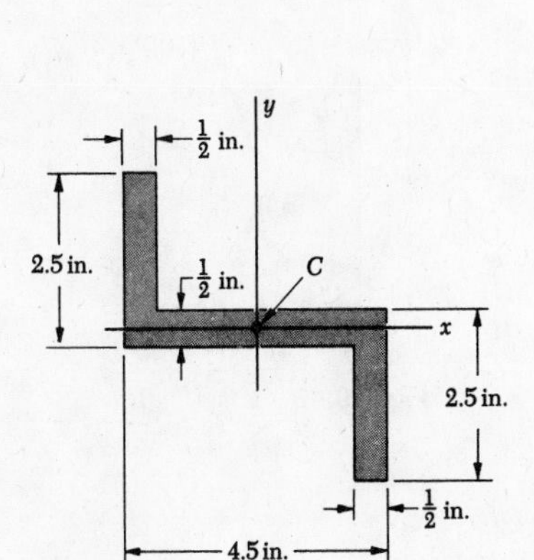

Fig. P9.38

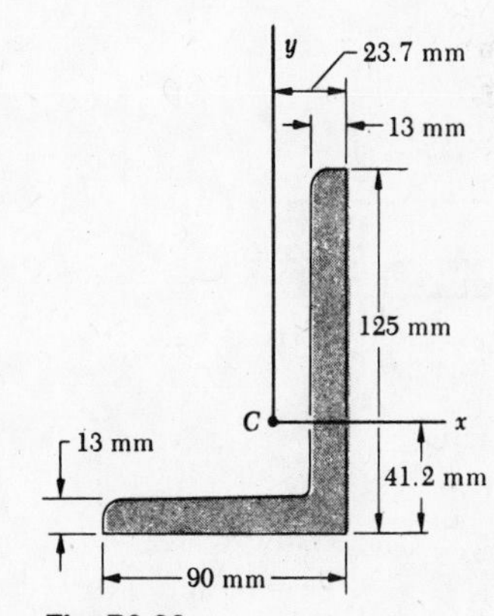

Fig. P9.39

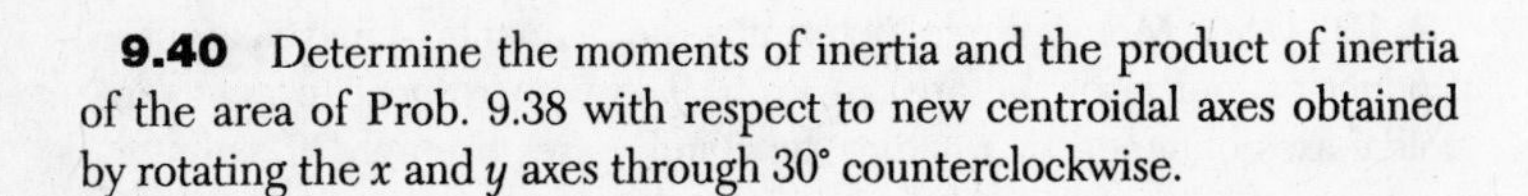

9.40 Determine the moments of inertia and the product of inertia of the area of Prob. 9.38 with respect to new centroidal axes obtained by rotating the x and y axes through 30° counterclockwise.

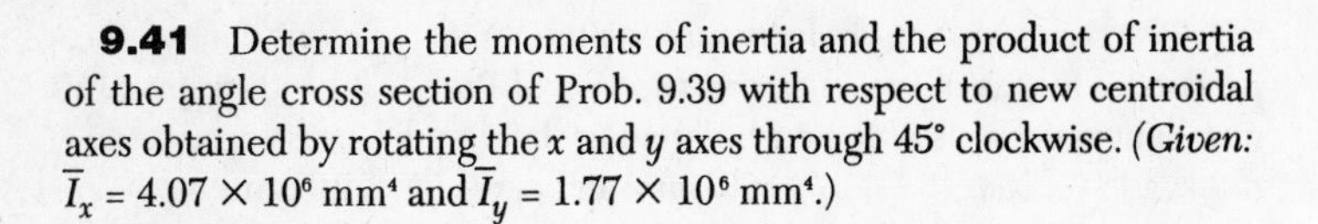

9.41 Determine the moments of inertia and the product of inertia of the angle cross section of Prob. 9.39 with respect to new centroidal axes obtained by rotating the x and y axes through 45° clockwise. (*Given:* $\bar{I}_x = 4.07 \times 10^6\ \text{mm}^4$ and $\bar{I}_y = 1.77 \times 10^6\ \text{mm}^4$.)

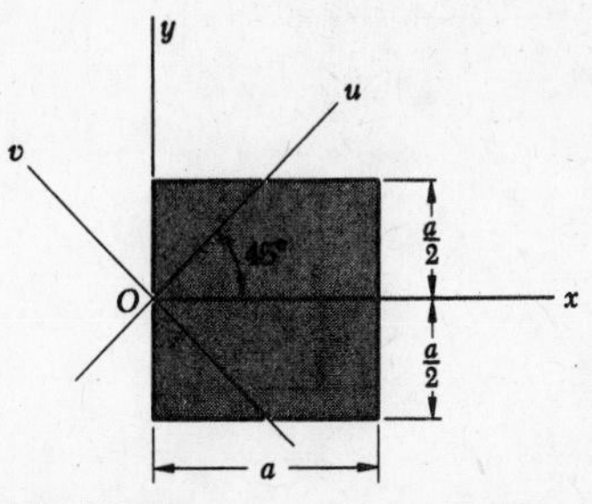

Fig. P9.42

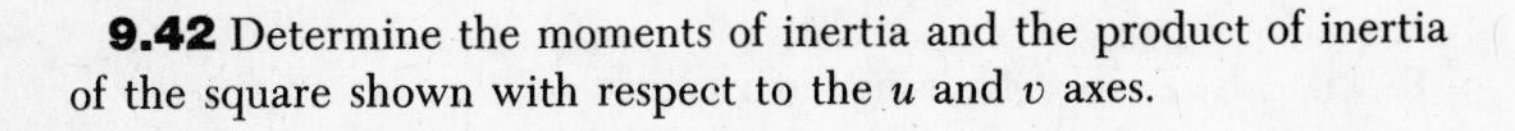

9.42 Determine the moments of inertia and the product of inertia of the square shown with respect to the u and v axes.

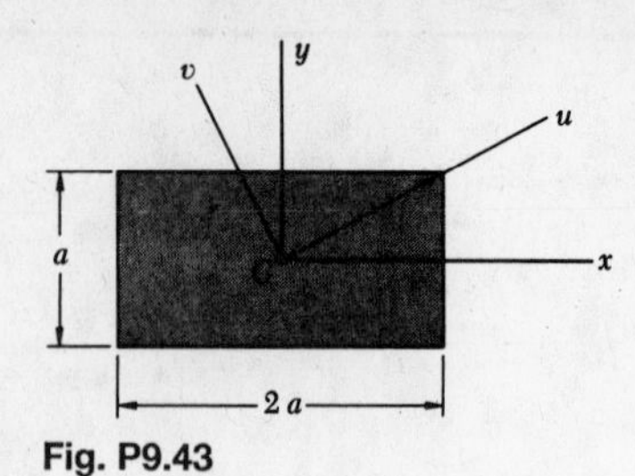

Fig. P9.43

9.43 Determine the moments of inertia and the product of inertia of the rectangle shown with respect to the u and v axes.

9.44 Determine the orientation of the principal axes through the centroid and the corresponding values of the moment of inertia for the area of Prob. 9.38.

9.45 Determine the orientation of the principal axes through the centroid and the corresponding values of the moment of inertia for the angle cross section of Prob. 9.39. (*Given:* $\bar{I}_x = 4.07 \times 10^6$ mm^4 and $\bar{I}_y = 1.77 \times 10^6$ mm^4.)

9.46 Determine the orientation of the principal axes through the centroid and the corresponding values of the moment of inertia for the angle cross section shown. Neglect the effect of fillets in computing $\bar{I}_{xy}$. (*Given:* $\bar{I}_x = 24.5$ in^4 and $\bar{I}_y = 8.7$ in^4.)

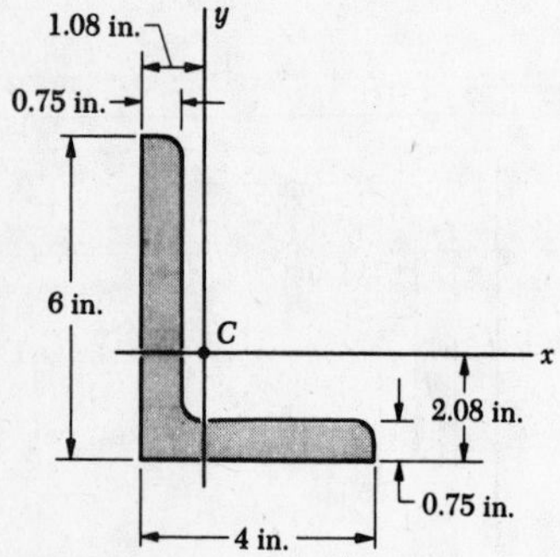

Fig. P9.46

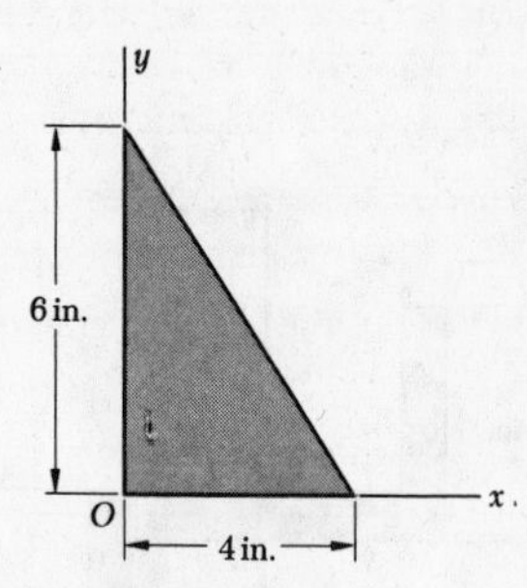

Fig. P9.47

9.47 Determine the orientation of the principal axes through O and the corresponding values of the moment of inertia for the triangle shown. (The product of inertia I_{xy} of the triangle is given in Sample Prob. 9.6.)

9.48 Using Mohr's circle, determine the moments of inertia and the product of inertia of the area of Prob. 9.38 with respect to new centroidal axes obtained by rotating the x and y axes through 30° counterclockwise.

9.49 Using Mohr's circle, determine the moments of inertia and the product of inertia of the angle cross section of Prob. 9.39 with respect to new centroidal axes obtained by rotating the x and y axes through 45° clockwise. (*Given:* $\bar{I}_x = 4.07 \times 10^6$ mm^4 and $\bar{I}_y = 1.77 \times 10^6$ mm^4.)

9.50 Solve Prob. 9.42, using Mohr's circle.

9.51 Solve Prob. 9.43, using Mohr's circle.

9.52 Using Mohr's circle, determine the orientation of the principal axes through the centroid and the corresponding values of the moment of inertia for the area of Prob. 9.38.

9.53 Using Mohr's circle, determine the orientation of the principal axes through the centroid and the corresponding values of the moment of inertia for the angle cross section of Prob. 9.39. (*Given:* $\bar{I}_x = 4.07 \times 10^6\ \text{mm}^4$ and $\bar{I}_y = 1.77 \times 10^6\ \text{mm}^4$.)

9.54 Solve Prob. 9.46, using Mohr's circle.

9.55 Solve Prob. 9.47, using Mohr's circle.

SECTIONS 9.11 to 9.15

9.56 A thin plate of mass m is cut in the shape of an isosceles triangle of width b and height h. Determine the mass moment of inertia of the plate with respect to (*a*) the centroidal axes AA' and BB' in the plane of the plate, (*b*) the centroidal axis CC' perpendicular to the plate.

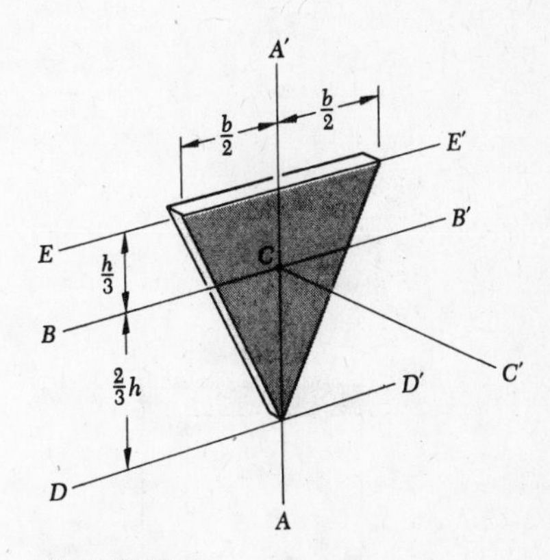

Fig. P9.56

9.57 Determine the mass moment of inertia of the plate of Prob. 9.56 with respect to (*a*) edge EE', (*b*) axis DD' which passes through a vertex and is parallel to edge EE'.

9.58 Determine the mass moment of inertia of a thin elliptical plate of mass m with respect to (*a*) the axes AA' and BB' of the ellipse, (*b*) the axis CC' perpendicular to the plate.

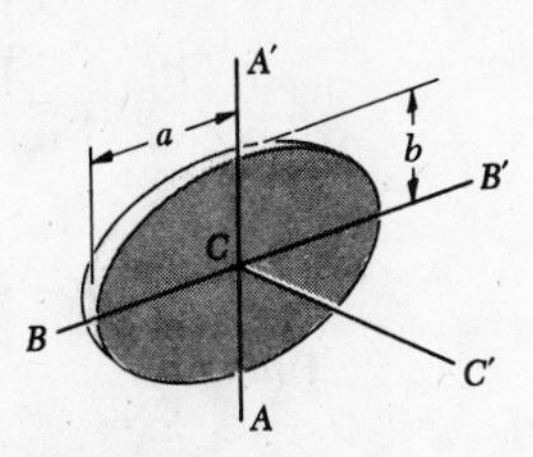

Fig. P9.58

9.59 Determine by direct integration the mass moment of inertia of the homogeneous regular tetrahedron of mass m and side a with respect to an axis through A perpendicular to the base BCD. (*Hint:* Use the result of part *b* of Prob. 9.56.)

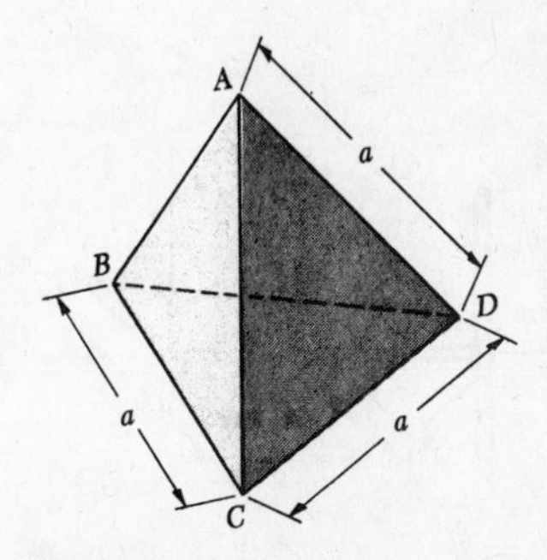

Fig. P9.59

9.60 Determine by direct integration the mass moment of inertia and the radius of gyration of a circular cylinder of radius a, length L, and uniform density ρ, with respect to a diameter of its base.

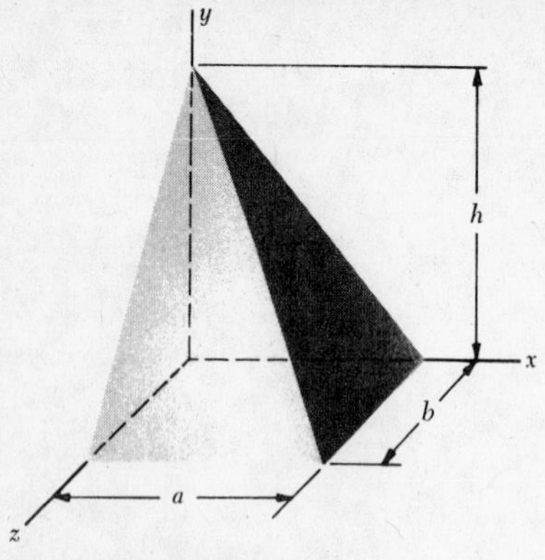

Fig. P9.61 and P9.62

9.61 Determine by direct integration the mass moment of inertia with respect to the x axis of the pyramid shown, assuming a uniform density and a mass m.

9.62 Determine by direct integration the mass moment of inertia with respect to the y axis of the pyramid shown, assuming a uniform density and a mass m.

9.63 The cross section of a small flywheel is shown. The rim and hub are connected by eight spokes (two of which are shown in the cross section). Each spoke has a cross-sectional area of 160 mm². Determine the mass moment of inertia and radius of gyration of the flywheel with respect to the axis of rotation. (Density of steel = 7850 kg/m³.)

9.64 Determine the mass moment of inertia and the radius of gyration of the steel flywheel shown with respect to the axis of rotation. The web of the flywheel consists of a solid plate 1 in. thick. (Specific weight of steel = 490 lb/ft³.)

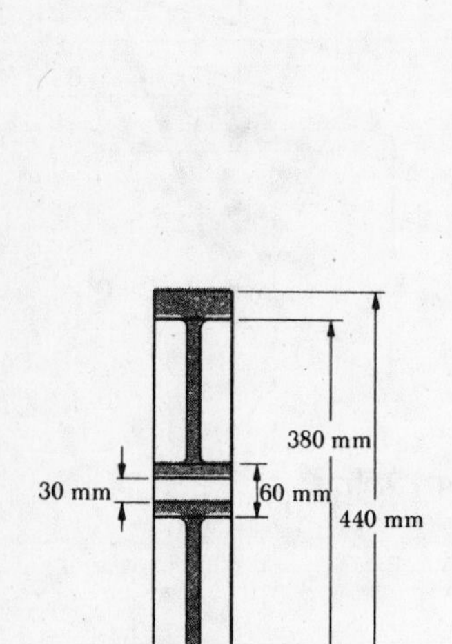

Fig. P9.63

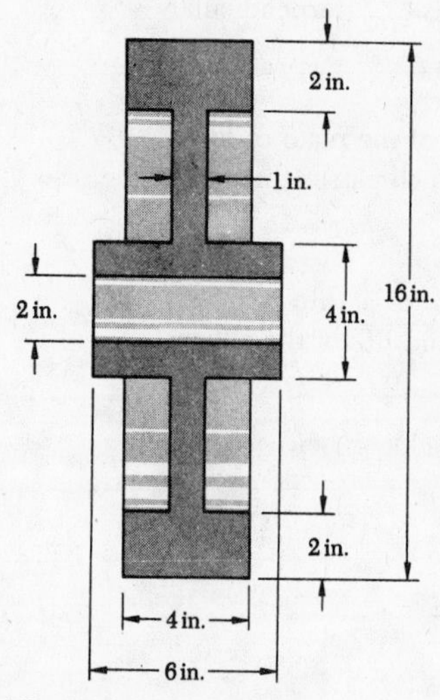

Fig. P9.64

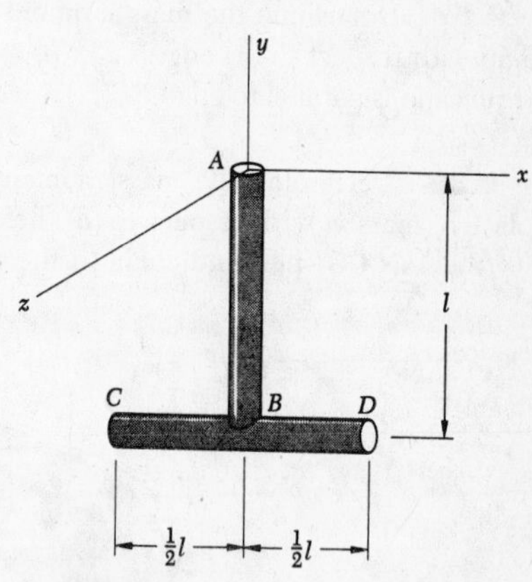

Fig. P9.65

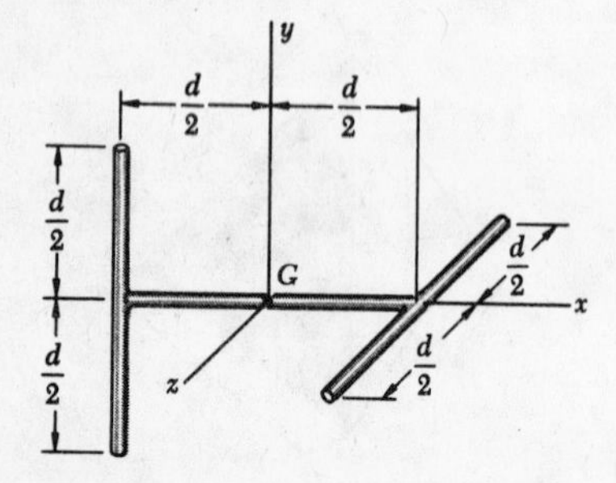

Fig. P9.66

9.65 Two rods, *each* of length l and mass m, are welded together as shown. Knowing that rod CD is parallel to the x axis, determine the mass moment of inertia of the composite body with respect to (a) the x axis, (b) the y axis, (c) the z axis.

9.66 Three slender homogeneous rods are welded together as shown. Denoting the mass of each rod by m, determine the mass moment of inertia and the radius of gyration of the assembly with respect to (a) the x axis, (b) the y axis, (c) the z axis.

9.67 and 9.68 Determine the mass moment of inertia and the radius of gyration of the steel machine element shown with respect to the x axis. (Specific weight of steel = 490 lb/ft^3; density of steel = 7850 kg/m^3.)

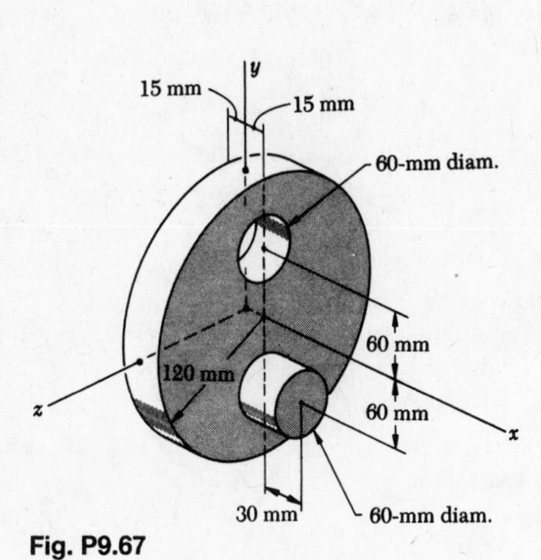

Fig. P9.67

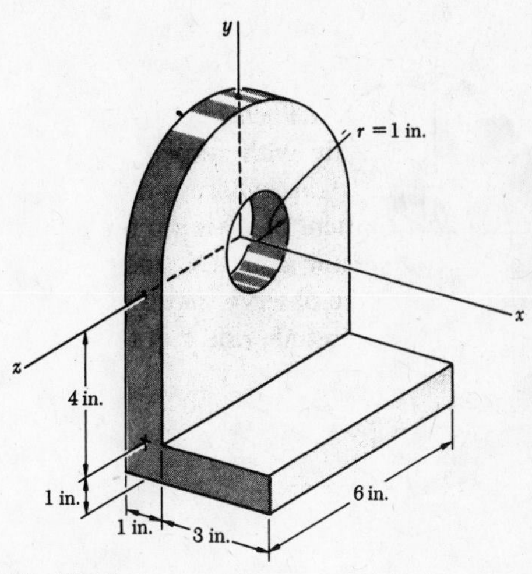

Fig. P9.68

9.69 A homogeneous wire, of weight 1.5 lb/ft, is used to form the figure shown. Determine the mass moment of inertia of the wire figure with respect to (*a*) the x axis, (*b*) the y axis, (*c*) the z axis.

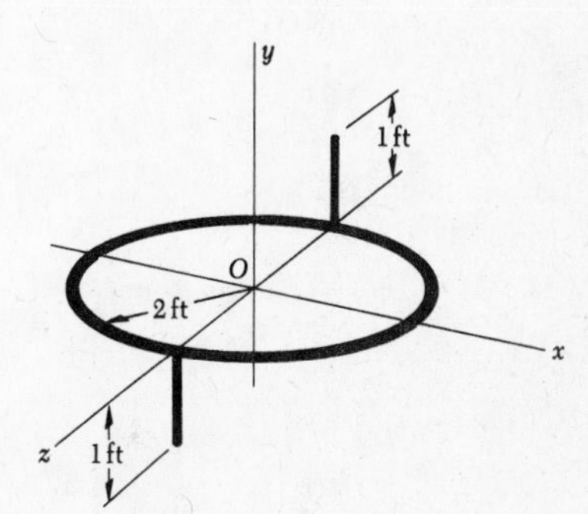

Fig. P9.69

9.70 A homogenous wire with a mass per unit length of 3 kg/m is used to form the figure shown. Determine the mass moment of inertia of the wire figure with respect to (*a*) the x axis, (*b*) the y axis, (*c*) the z axis.

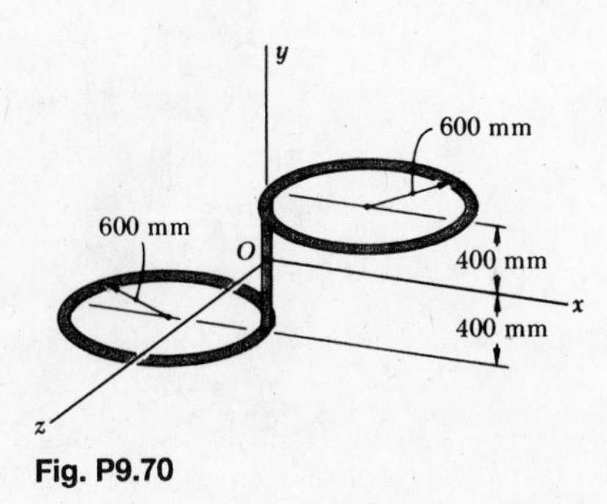

Fig. P9.70

9.71 Two holes, each of diameter 50 mm, are drilled through the steel block shown. Determine the mass moment of inertia of the body with respect to the axis of either of the holes. (Density of steel = 7850 kg/m^3.)

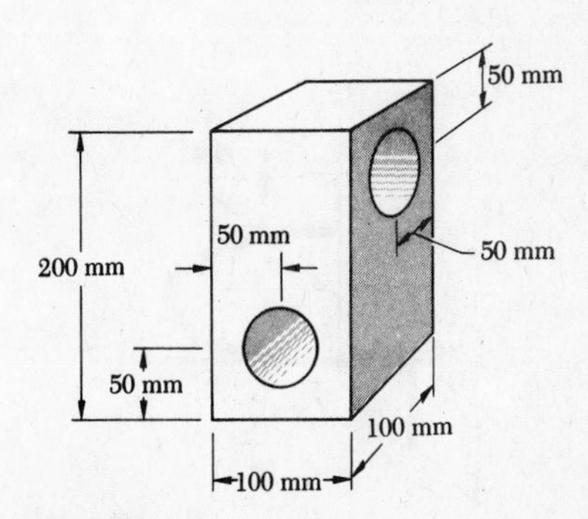

Fig. P9.71

SECTIONS 9.16 and 9.17

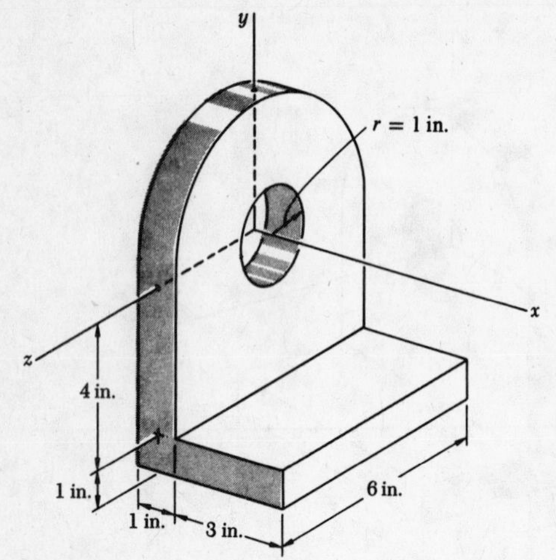

Fig. P9.72

9.72 through 9.74 Determine the mass products of inertia I_{xy}, I_{yz}, and I_{zx} of the steel machine element shown. (Specific weight of steel = 490 lb/ft³; density of steel = 7850 kg/m³.)

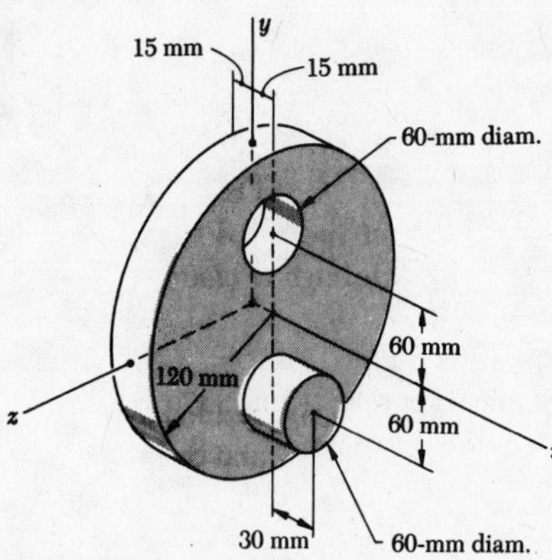

Fig. P9.73

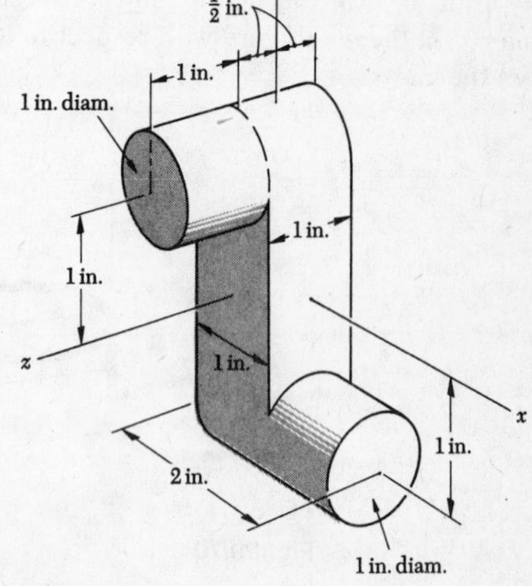

Fig. P9.74

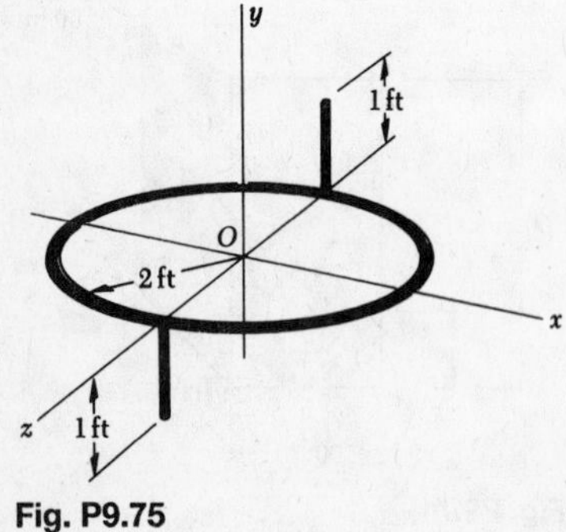

Fig. P9.75

9.75 A homogeneous wire, of weight 1.5 lb/ft, is used to form the figure shown. Determine the mass products of inertia I_{xy}, I_{yz}, and I_{zx} of the wire figure.

9.76 A homogeneous wire with a mass per unit length of 3 kg/m is used to form the figure shown. Determine the mass products of inertia I_{xy}, I_{yz}, and I_{zx} of the wire figure.

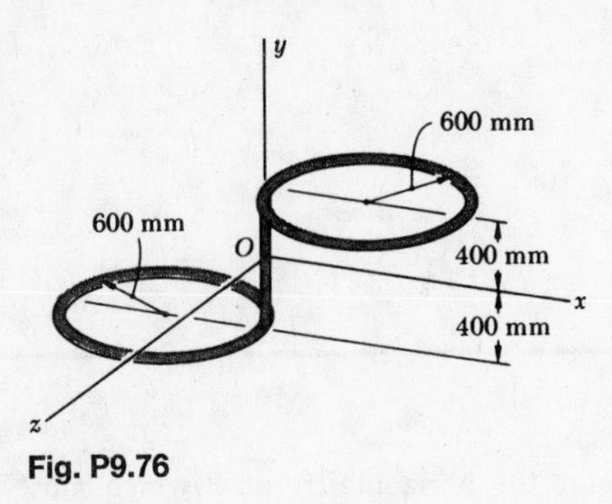

Fig. P9.76

9.77 A section of sheet steel, 2 mm thick, is cut and bent into the machine component shown. Knowing that the density of steel is 7850 kg/m³, determine the mass products of inertia I_{xy}, I_{yz}, and I_{zx} of the component.

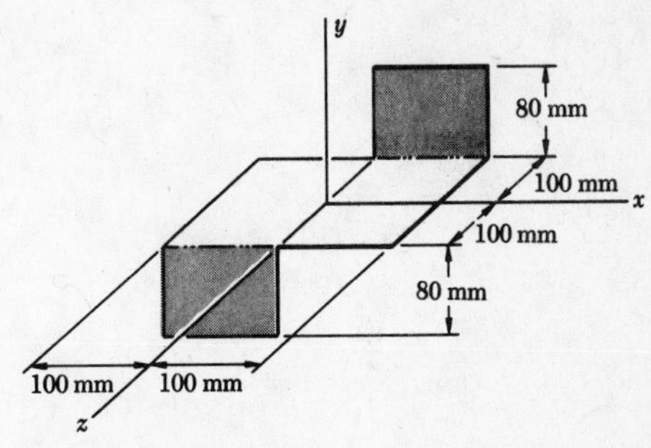

Fig. P9.77

9.78 Determine the mass moment of inertia of the bent wire of Probs. 9.70 and 9.76 with respect to the axis through O which forms equal angles with the x, y, and z axes.

9.79 Determine the mass moment of inertia of the bent wire of Probs. 9.69 and 9.75 with respect to the axis through O which forms equal angles with the x, y, and z axes.

9.80 Determine the mass moment of inertia of the forging of Sample Prob. 9.12 with respect to an axis through O characterized by the unit vector $\boldsymbol{\lambda} = 0.5\mathbf{i} + 0.5\mathbf{j} + 0.707\mathbf{k}$.

9.81 Consider a rectangular prism of mass m and sides a, b, and c such that $a > b > c$. What is (*a*) the largest, (*b*) the smallest mass moment of inertia with respect to any line through the centroid of the prism?

9.82 (*a*) Show that the ellipsoid of inertia at the center of a regular tetrahedron is a sphere. (*b*) Using this fact and the result of Prob. 9.59, determine the principal moments of inertia of the tetrahedron at one of its corners.

9.83 Using the result of Prob. 9.82, determine the mass moment of inertia of a homogeneous regular tetrahedron with respect to one of its edges.

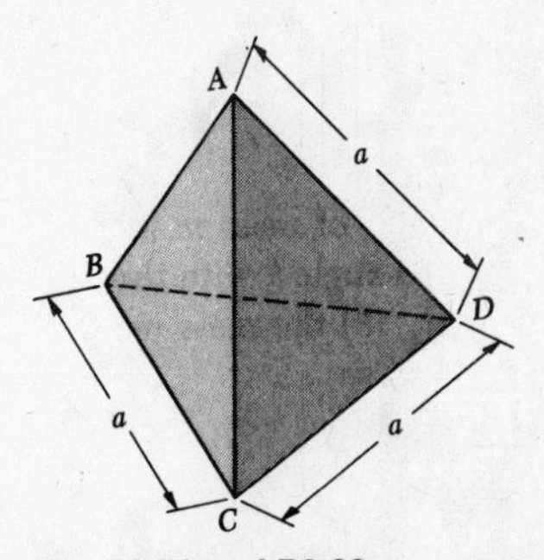

Fig. P9.82 and P9.83

CHAPTER 10
VIRTUAL WORK

SECTIONS 10.1 to 10.5

10.1 Determine the horizontal force **P** which must be applied at A to maintain the equilibrium of the linkage.

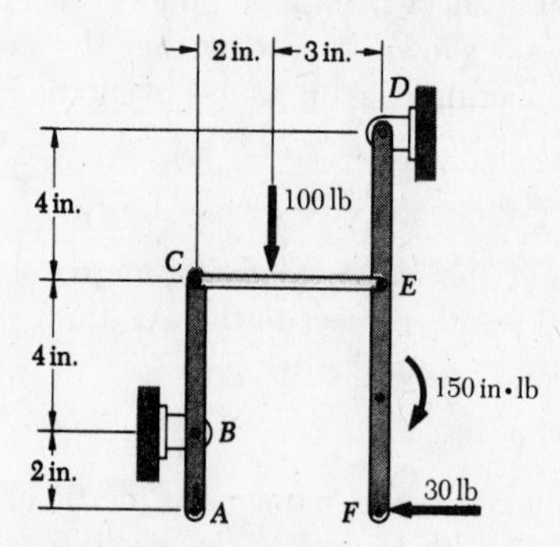

Fig. P10.1 and P10.2

10.2 Determine the couple **M** which must be applied to member AC to maintain the equilibrium of the linkage.

10.3 Determine the magnitude of the force **P** required to maintain the equilibrium of the linkage shown.

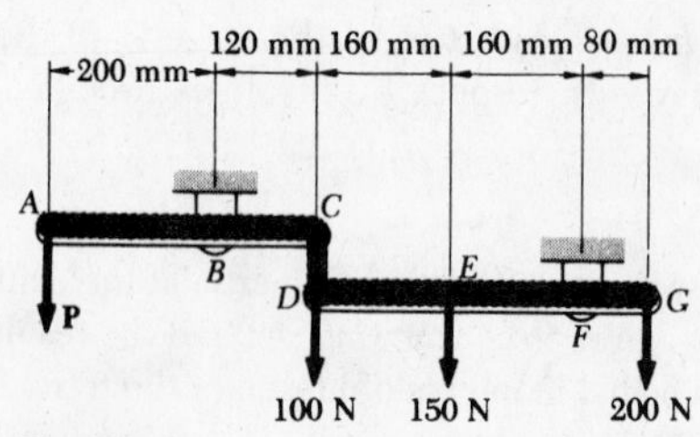

Fig. P10.3

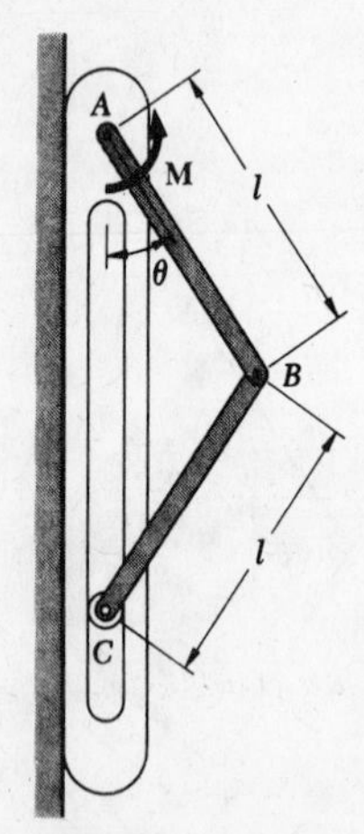

Fig. P10.4

10.4 Each of the uniform rods AB and BC is of weight W. Derive an expression for the magnitude of the couple **M** required to maintain equilibrium.

10.5 Derive an expression for the magnitude of the couple **M** required to maintain equilibrium.

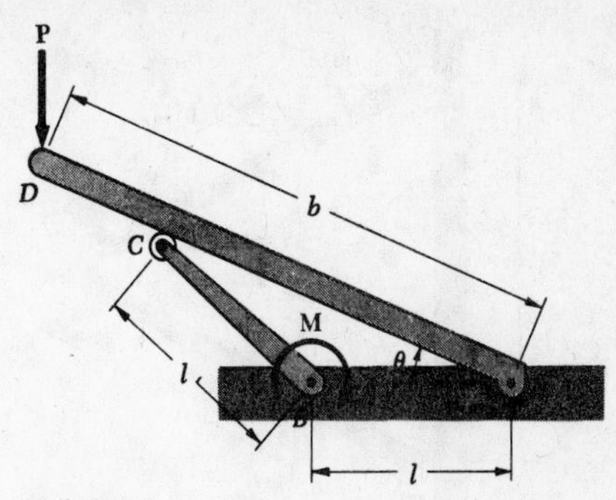

Fig. P10.5

10.6 The slender rod AB is attached to a collar at A and rests on a wheel at C. Neglecting the effect of friction, derive an expression for the magnitude of the force **Q** required to maintain equilibrium.

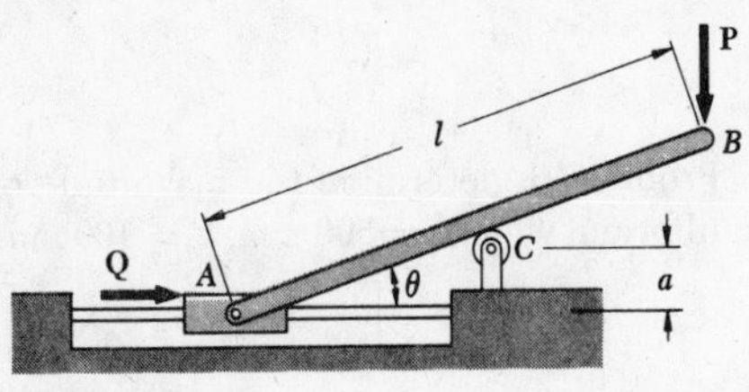

Fig. P10.6

10.7 Three links, each of length l, are connected as shown. Knowing that the line of action of the force **Q** passes through point A, derive an expression for the magnitude of **Q** required to maintain equilibrium.

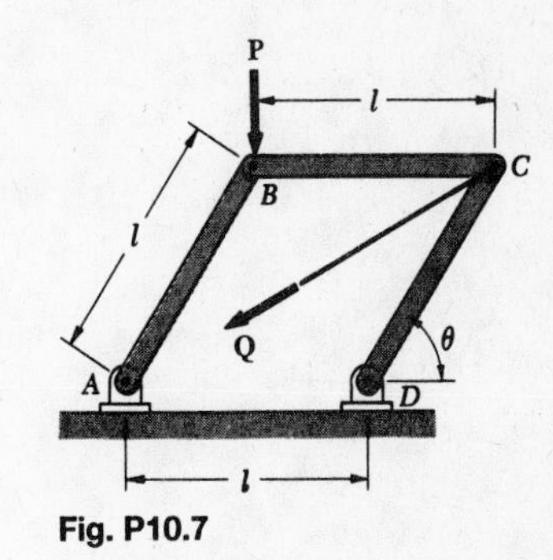

Fig. P10.7

10.8 Derive an expression for the magnitude of the couple **M** required to maintain equilibrium.

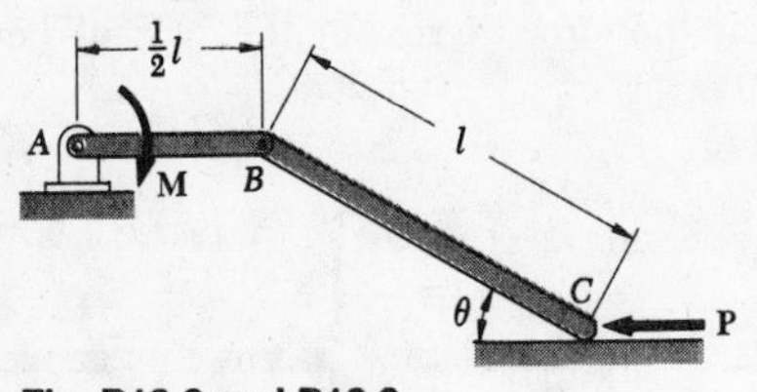

Fig. P10.8 and P10.9

10.9 Denoting by μ the coefficient of friction between end C of rod BC and the horizontal surface, determine the largest magnitude of the couple **M** for which equilibrium is maintained.

Fig. P10.10

10.10 A load **W** of magnitude 800 N is applied to the mechanism at B. Neglecting the weight of the mechanism, determine the value of θ corresponding to equilibrium. The constant of the spring is $k = 10$ kN/m, and the spring is unstretched when bar AC is horizontal.

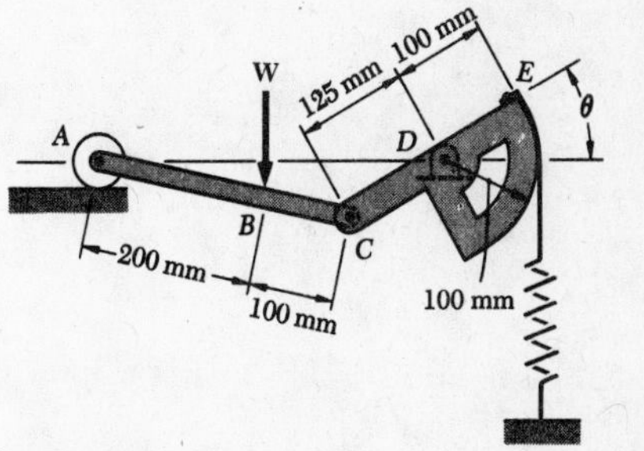

10.11 Denoting by μ the coefficient of friction between the collar C and the horizontal rod, determine the largest magnitude of the couple **M** for which equilibrium is maintained. Explain what happens if $\mu \geq \tan\theta$.

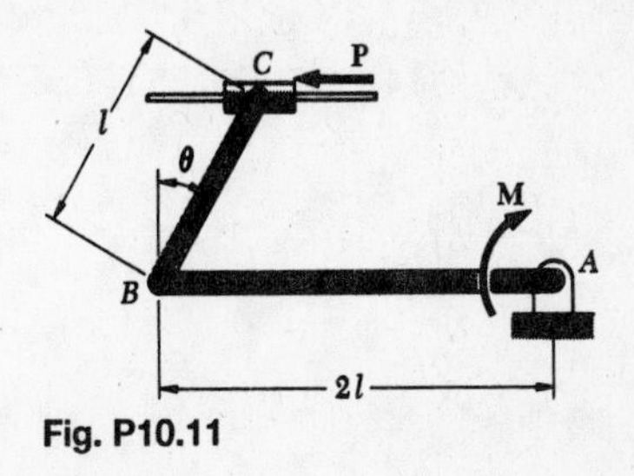

Fig. P10.11

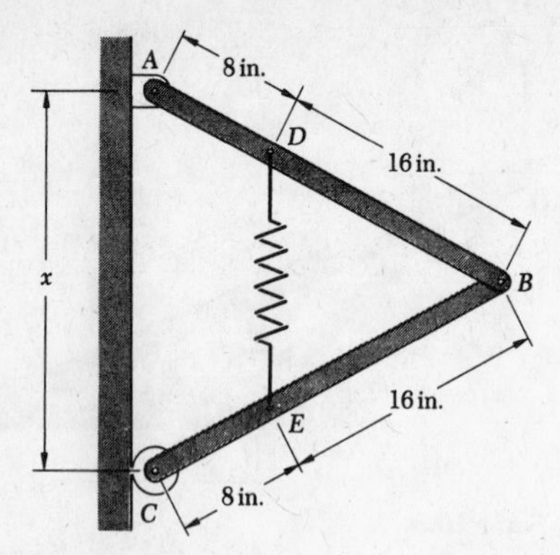

Fig. P10.12

10.12 Two 10-lb bars AB and BC are connected by a pin at B and, by a spring DE. When unstretched, the spring is 6 in. long; the constant of the spring is 5 lb/in. Determine the value of x corresponding to equilibrium.

10.13 In Prob. 10.5, determine the magnitude of the couple $\mathbf{M}$ required for equilibrium when $l = 80$ mm, $b = 200$ mm, $P = 300$ N, and $\theta = 30°$.

10.14 In Prob. 10.6, determine the magnitude of the force $\mathbf{Q}$ required for equilibrium when $l = 500$ mm, $a = 100$ mm, $\theta = 30°$ and $P = 500$ N.

10.15 Determine the value of θ corresponding to the equilibrium position of the mechanism of Prob. 10.5 when $P = 400$ N, $M = 50$ N • m, $b = 300$ mm, and $l = 90$ mm.

10.16 Determine the value of θ corresponding to the equilibrium position of the mechanism of Prob. 10.6 when $P = 500$ N, $Q = 400$ N, $a = 100$ mm, and $l = 500$ mm.

10.17 The elevation of the overhead platform is controlled by two identical mechanisms, only one of which is shown. A load of 5 kN is applied to the mechanism shown. Neglecting the weight of the mechanism, determine the force exerted by the hydraulic cylinder on pin E when $\theta = 30°$.

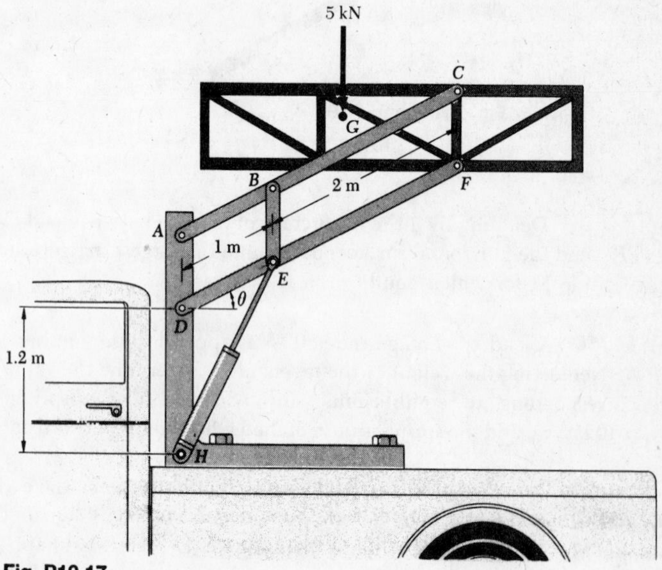

Fig. P10.17

10.18 For the mechanism of Prob. 10.17, (*a*) express the force exerted by the hydraulic cylinder on pin E as a function of the length HE, (*b*) determine the largest possible value of the angle θ if the maximum force that the cylinder can exert on pin E is 25 kN.

10.19 The top surface of the work platform is maintained at a given elevation by the single hydraulic cylinder shown. Knowing that members AF and BD are each of length l, determine the magnitude of the force exerted by the hydraulic cylinder when $\theta = 30°$.

10.20 Solve Prob. 10.19, assuming that the hydraulic cylinder has been moved and attached to points E and H.

10.21 A slender rod of length l is attached to a collar at B and rests on a frictionless circular cylinder of radius r. Knowing that the collar may slide freely along a vertical guide, derive an expression for the magnitude of the force **Q** required to maintain equilibrium.

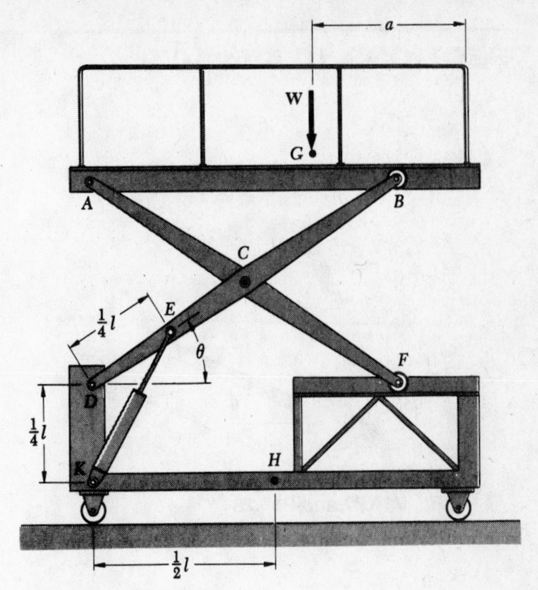

Fig. P10.19

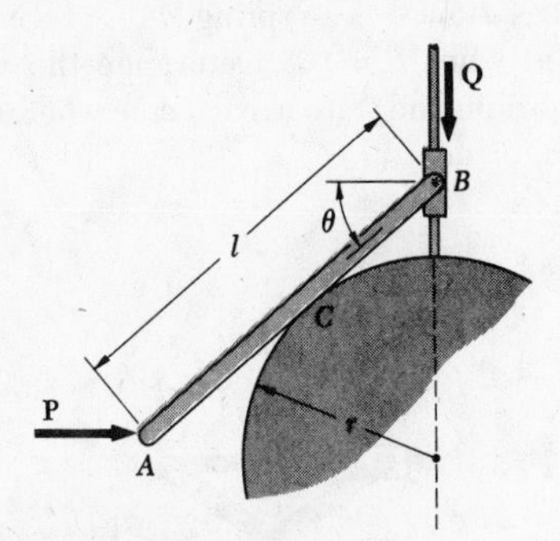

Fig. P10.21

10.22 Solve Prob. 10.21, assuming that the force **P** is removed and that a couple **M**, directed counterclockwise, is applied to rod AB.

10.23 If gripping forces of magnitude $Q = 450$ lb are desired, determine the magnitude P of the forces which must be applied to the pliers. Also show that the required magnitude P is independent of the position of the object gripped by the jaws.

10.24 Determine the weight W which balances the 45-lb load.

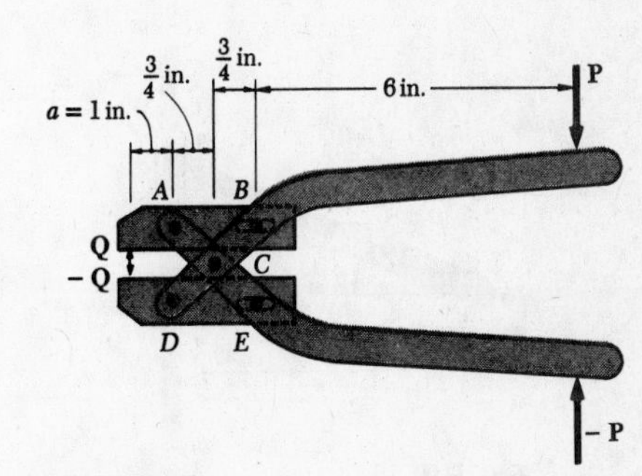

Fig. P10.23

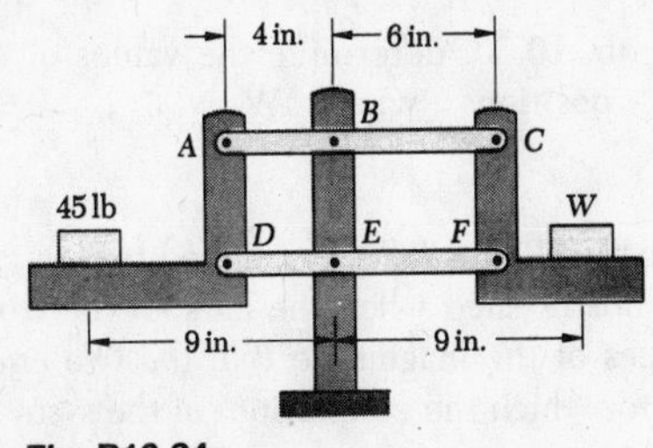

Fig. P10.24

10.25 Using the method of Sec. 10.8, solve Prob. 10.10.

10.26 Using the method of Sec. 10.8, solve Prob. 10.12.

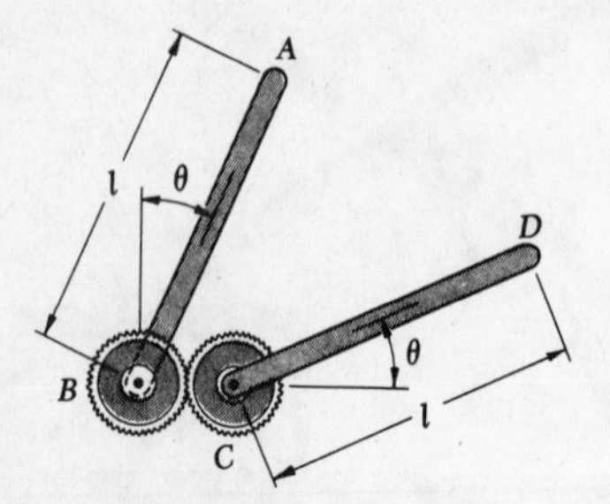

Fig. P10.27 and P10.28

10.27 Two uniform rods, each of weight W, are attached to gears of equal radii as shown. Determine the positions of equilibrium of the system and state in each case whether the equilibrium is stable, unstable, or neutral.

10.28 A vertical force **P** of magnitude 10 lb is applied to rod CD at D. Knowing that the uniform rods AB and CD weigh 5 lb each, determine the positions of equilibrium of the system and state in each case whether the equilibrium is stable, unstable, or neutral.

10.29 Two rods AF and BD, each of length 15 in., and a third rod BE are connected as shown. The spring EA is of constant $k = 9$ lb/in. and is undeformed when $\theta = 0$. Determine the values of θ corresponding to equilibrium and state in each case whether the equilibrium is stable, unstable, or neutral.

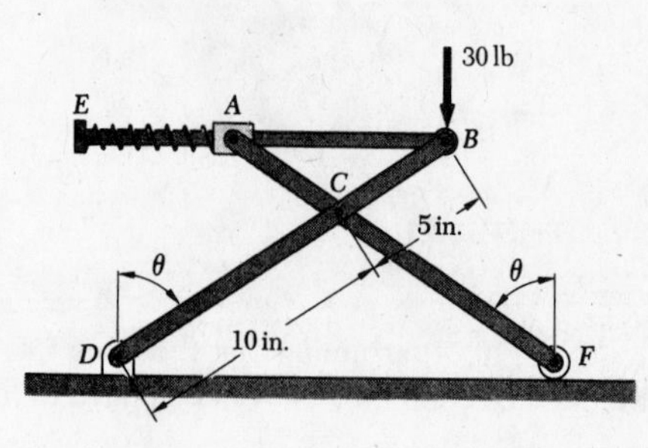

Fig. P10.29

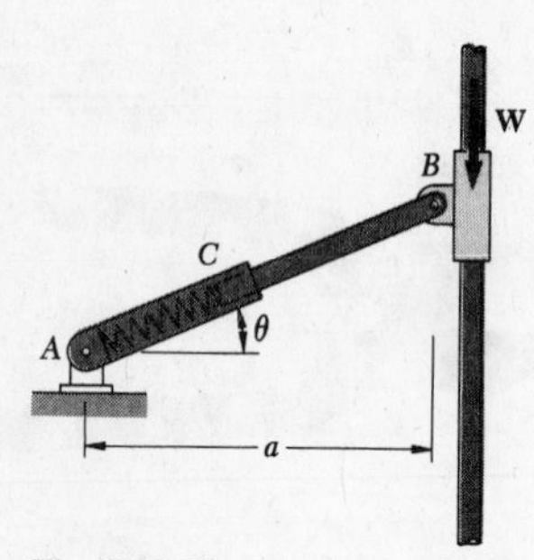

Fig. P10.30

10.30 The internal spring AC is of constant k and is undeformed when $\theta = 45°$. Derive an equation defining the values of θ corresponding to equilibrium positions.

10.31 In Prob. 10.30, determine the values of θ corresponding to equilibrium positions when $W = 75$ N, $k = 2$ kN/m, and $a = 500$ mm.

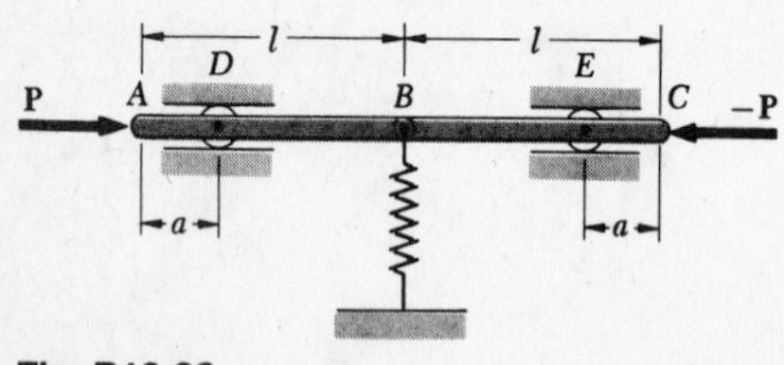

Fig. P10.32

10.32 Two bars AB and BC are attached to a single spring of constant k which is unstretched when the bars are horizontal. Determine the range of values of the magnitude P of the two equal and opposite forces **P** and **–P** for which the equilibrium of the system is stable in the position shown.

10.33 A vertical load $\mathbf{W}$ is applied to the linkage at B. The constant of the spring is k, and the spring is unstretched when AB and BC are horizontal. Neglecting the weight of the linkage, derive an equation in θ, W, l, and k, which must be satisfied when the linkage is in equilibrium.

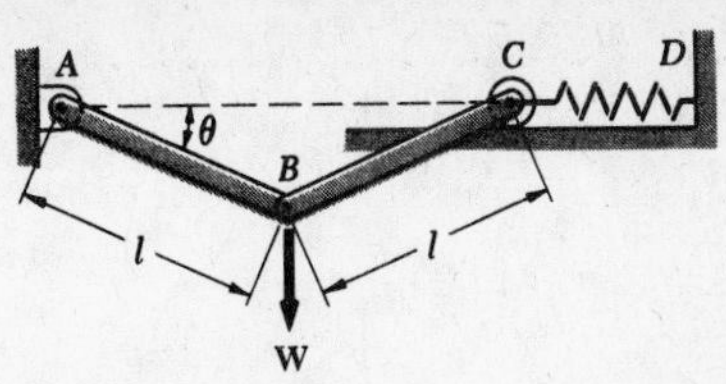

Fig. P10.33 and P10.34

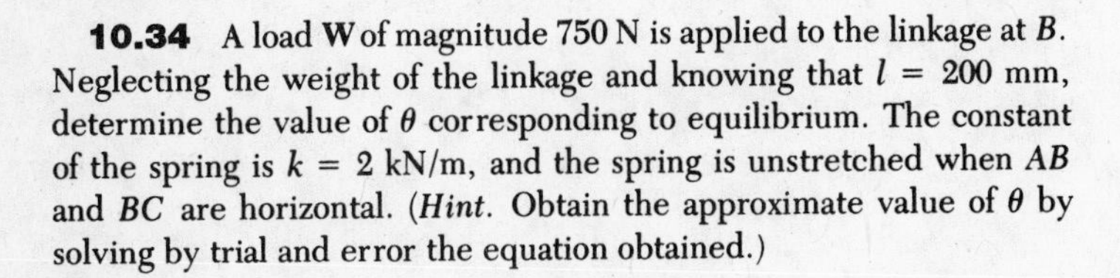

10.34 A load $\mathbf{W}$ of magnitude 750 N is applied to the linkage at B. Neglecting the weight of the linkage and knowing that $l = 200$ mm, determine the value of θ corresponding to equilibrium. The constant of the spring is $k = 2$ kN/m, and the spring is unstretched when AB and BC are horizontal. (*Hint.* Obtain the approximate value of θ by solving by trial and error the equation obtained.)

10.35 (*a*) Derive an expression for the moment of the couple $\mathbf{M}$ required to maintain equilibrium. (*b*) Determine the value of θ corresponding to the equilibrium position when $P = 20$ lb, $M = 80$ lb·in., $l = 4$ in., and $b = 10$ in., and state whether the equilibrium is stable, unstable, or neutral.

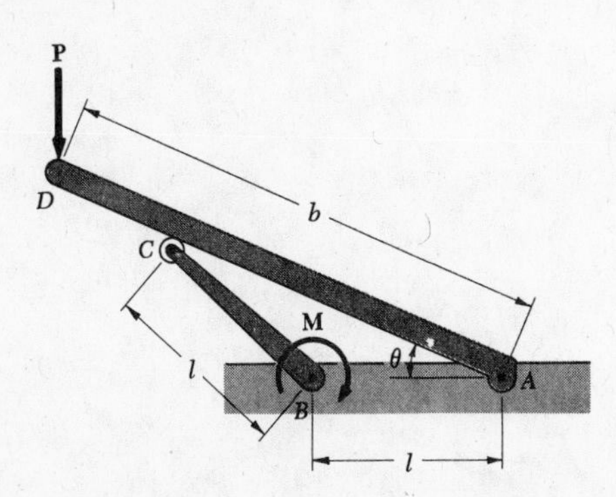

Fig. P10.35

10.36 Members ACE and DCB are each of length 25 in. and are connected by a pin at their midpoints C. A load $\mathbf{P}$ of magnitude 200 lb is applied to member DF. If $h = 15$ in. and $a = 30$ in., determine (*a*) the tension in the spring AD, (*b*) the unstretched length of the spring, knowing that the spring constant is 40 lb/in.

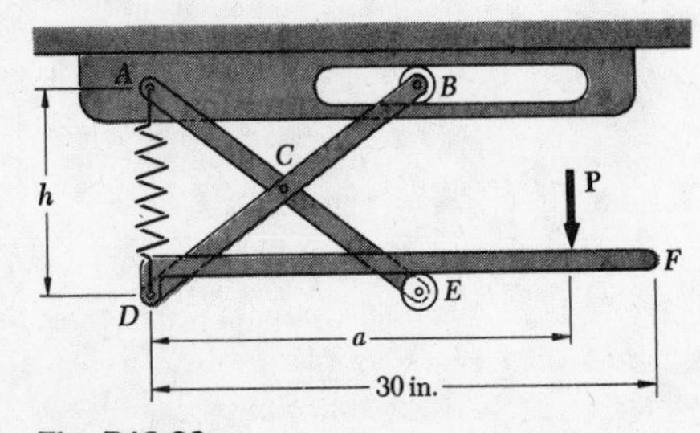

Fig. P10.36

ANSWERS TO PROBLEMS

CHAPTER 2

2.1 3240 N ∠40.9°.
2.2 585 lb ∠23.0°.
2.3 4000 lb ∠9.2°.
2.4 14.3 kN ∠19.9°.
2.5 32.4°.
2.6 123.4°.
2.7 $P = 410$ N; $R = 919$ N.
2.8 14.73 lb; 30.2 lb.
2.9 394 N ∠58.6°.
2.10 43.6 lb ∠78.4°.
2.12 (*a*) 90°. (*b*) 205 N.
2.13 −2.35 kN, +0.855 kN.
2.14 69.3 lb; 34.6 lb.
2.15 25 lb→, 60 lb↓.
2.16 1200 N; 1039 N.
2.17 (40 lb) 20 lb→, 34.6 lb↑;
(50 lb) 47.0 lb→, 17.10 lb↑;
(60 lb) 38.6 lb→, 46.0 lb↓.
2.18 (80 N) +69.3 N, −40.0 N;
(100 N) +34.2 N, −94.0 N;
(120 N) −91.9 N, −77.1 N.
2.19 (80 N) 51.4 N↗, 61.3 N↘;
(120 N) 112.8 N↙, 41.0 N↘.
2.20 (40 lb) 30.6 lb↗, 25.7 lb↖;
(60 lb) 20.5 lb↗, 56.4 lb↘.
2.21 211 N ∠86.9°.
2.22 105.6 lb ∠3.1°.
2.25 14.04 kips.
2.26 6.3° and 133.7°.
2.27 (*a*) 48.2°. (*b*) Impossible.
2.28 15.50 lb ∠21.6°.
2.29 $T_{AC} = 120.1$ lb; $T_{BC} = 156.3$ lb.
2.30 $T_{AC} = 981$ N; $T_{BC} = 2350$ N.
2.31 $T_{AC} = 716$ N; $T_{BC} = 381$ N.
2.32 $T_{AC} = 326$ lb; $T_{BC} = 265$ lb.
2.33 $F = 2.87$ kN; $\alpha = 75°$.
2.34 (*a*) 30°. (*b*) $T_{AC} = 300$ lb; $T_{BC} = 520$ lb.
2.35 (*a*) 60°. (*b*) $T_{AB} = T_{BC} = 346$ lb.
2.36 913 N ∠82.5°.
2.37 $A = 9.81$ kN; $B = 4.62$ kN.
2.38 $C = 2.17$ kN; $D = 3.75$ kN.
2.39 $T_{AC} = 46.2$ lb; $T_{BC} = 36.9$ lb.
2.40 65.2 lb $< P <$ 150 lb.
2.41 (*a*) 2.45 kN. (*b*) 1.838 kN.
2.42 1.252 in.
2.43 (*a*) 18 lb. (*b*) 24 lb.
2.44 (*a*) 1396 N. (*b*) 238 N.
2.45 (*a*) +113.3 N, +217 N, −52.8 N.
(*b*) 63.1°, 30.0°, 102.2°.
2.46 (*a*) +65.9 N, +229.8 N, +181.2 N.
(*b*) 77.3°, 40.0°, 52.8°.
2.47 (*a*) −78.6 lb, +282 lb, −66.0 lb.
(*b*) 105.2°, 20.0°, 102.7°.
2.48 (*a*) +78.6 lb, +282 lb, −66.0 lb.
(*b*) 74.8°, 20.0°, 102.7°.
2.49 1444 N; 61.0°, 124.6°, 48.3°.
2.50 721 lb; 109.4°, 116.3°, 33.7°.
2.51 61.0°; +105.7 lb, +191.5 lb, +121.2 lb.
2.52 48.4 N; 34.3°.
2.53 −400 N, +400 N, −700 N.
2.54 −300 N, +300 N, +150 N.
2.55 116.4°, 63.6°, 141.1°.
2.56 131.8°, 48.2°, 70.5°.
2.57 +1500 lb, −1500 lb, +600 lb.
2.58 (*a*) 54.7°, 125.3°. (*b*) 60°, 120°.
2.59 $T_{AC} = 21$ kN; $T_{AD} = 64.3$ kN.

2.60 $T_{AB} = 52.0$ kN; $T_{AD} = 85.7$ kN.
2.61 $T_{DA} = 2.00$ kN; $T_{DB} = 3.50$ kN; $T_{DC} = 1.500$ kN.
2.62 $T_{AD} = 1.750$ kN, $T_{BD} = 1.750$ kN, $T_{CD} = 3.00$ kN.
2.63 2775 lb.
2.64 888 lb.
2.65 $T_{AD} = 262.5$ lb; $T_{BD} = 195.0$ lb; $T_{CD} = 187.5$ lb.
2.66 4.80 kN, down.
2.67 $T_{AD} = 11.90$ kN; $T_{BD} = 16.10$ kN; $T_{CD} = 4.00$ kN.
2.68 $T_{AD} = 7.21$ lb; $T_{BD} = T_{CD} = 6.50$ lb.
2.69 $T_{AD} = 139.9$ N, $T_{BD} = T_{CD} = 115.7$ N.
2.70 $P = 101.8$ N; $T_{AB} = 211$ N.
2.71 (*a*) 6.3 lb. (*b*) 7.2 lb.
2.72 (*a*) 12.6 lb. (*b*) 50.4 lb.
2.73 $T_{AD} = 103.5$ N; $T_{BD} = 51.8$ N; $T_{CD} = 89.7$ N.
2.74 $T_{AD} = 8.50$ kN; $T_{BD} = 19.50$ kN; $T_{CD} = 14.00$ kN.
2.75 $T_{AC} = 1192$ lb; $T_{BC} = 898$ lb.
2.76 (*a*) 50.0°. (*b*) $T_{AC} = 1554$ lb; $T_{BC} = 688$ lb.

CHAPTER 3

3.1 32.5 N · m↻.
3.2 37.5 N · m↻; $\alpha = 20°$.
3.3 (*a*) 88.8 N · m↻. (*b*) 237 N ∠53.1°.
3.4 (*a*) 88.8 N · m↻. (*b*) 395 N←. (*c*) 280 N ∠45°.
3.5 241 lb · in.↺.
3.6 167.0 lb · in.↺.
3.7 233 lb · in.↻.
3.8 547 lb · in.↻.
3.9 2.33 in.
3.10 (*a*) 28**i** + 22**j** − 23**k**. (*b*) 11**i** − 22**k**. (*c*) **i** + 2**j** + **k**.
3.11 (*a*) −23**i** − 11**j** + 2**k**. (*b*) −30**j** + 18**k**. (*c*) 0.
3.12 (*a*) −(7200 lb · ft)**i** −(1600 lb · ft)**j** + (3200 lb · ft)**k**. (*b*) (5600 lb · ft)**j** + (3200 lb · ft)**k**.
3.13 (*a*) −(1200 lb · ft)**j** + (2400 lb · ft)**k**. (*b*) (5400 lb · ft)**i** + (4200 lb · ft)**j** + (2400 lb · ft)**k**.
3.14 (*a*) (24 N · m)**i** − (24 N · m)**k**. (*b*) (5.40 N · m)**i** − (18 N · m)**j** − (14.40 N · m)**k**.
3.15 (*a*) (40 N · m)**i** − (90 N · m)**j** − (40 N · m)**k**. (*b*) (9 N · m)**i** − (86.25 N · m)**j** − (42 N · m)**k**.
3.16 −(2160 lb · in.)**i** + (4320 lb · in.)**j** + (360 lb · in.)**k**.
3.17 (36 N · m)**i** + (24 N · m)**j** + (32 N · m)**k**.
3.18 269 mm.
3.19 11.53 in.
3.20 **P** · **Q** = 0; **P** · **S** = −11; **Q** · **S** = 2.
3.21 **P** · **Q** = 0; **P** · **S** = −15; **Q** · **S** = 24.
3.22 (*a*) 46.0°. (*b*) 68.9°.
3.23 590 lb.
3.24 (*a*) 15.0°. (*b*) 290 N.
3.25 (*a*) 23.5°. (*b*) 413 N.
3.26 (*a*) 59.05°. (*b*) 720 N.
3.27 (*a*) 70.5°. (*b*) 300 N.
3.28 (*a*) 44.0°. (*b*) 395 lb.
3.29 (*a*) 91.9°. (*b*) −18.18 lb.
3.30 −1; −1; +1.
3.31 −6.
3.32 $M_x = 24$ kN · m, $M_y = -16$ kN · m, $M_z = -38.4$ kN · m.
3.33 4.88 kN.
3.35 −280 lb · in.
3.36 (*a*) −299 lb · in. (*b*) +212 lb · in.
3.37 (*a*) +144 lb · in. (*b*) +127.1 lb · in.
3.38 +38.6 N · m.
3.39 +124.2 N · m.
3.40 136.9 mm.
3.41 229 mm.
3.42 8.50 in.
3.43 7.06 in.
3.45 (*a*) 271 N. (*b*) 390 N. (*c*) 250 N.
3.46 280 lb · in.↻.
3.47 (*a*) 20 lb. (*b*) 16 lb. (*c*) 12 lb.
3.48 $M = 618$ N · m; $\theta_x = 104.0°$, $\theta_y = 90.0°$, $\theta_z = 14.0°$.
3.49 $M = 200$ lb · ft; $\theta_x = 126.9°$, $\theta_y = 90.0°$, $\theta_z = 36.9°$.
3.50 0.555M**i** + 1.279M**j** + 0.894M**k**.

3.51 $M_x = 3.6$ kN · m, $M_y = 7.72$ kN · m, $M_z = 0$.
3.52 **F** = 16 kips ↓, **M** = 192 kip · in.↺.
3.53 14.5°.
3.54 (*a*) 960 N ∠60°, 28.9 mm to right of *O*. (*b*) 960 N ∠60°, 50 mm below *O*.
3.55 **B** = 4**P**; **C** = −3**P**.
3.56 (*a*) **F** = 50 lb ∖60°, $\mathbf{M}_A$ = 308 lb · in.↺. (*b*) 102.7 lb← at *A* and 102.7 lb→ at *B*.
3.57 (*a*) **F** = 50 lb ∖60°, **M** = 233 lb · in.↺. (*b*) **A** = 77.7 lb←; **B** = 77.7 lb→.
3.58 **F** = (4 kips)**i**, $\mathbf{M}_G$ = −(3.18 kip · in.)**j** − (16 kip · in.)**k**.
3.59 **F** = −(100 kN)**j**, $\mathbf{M}_G$ = −(5 kN · m)**i** − (12.5 kN · m)**k**.
3.60 **F** = −(173.2 N)**j** + (100 N)**k**, **M** = (7.5 N · m)**i** − (6 N · m)**j** − (10.39 N · m)**k**.
3.61 (*a*) **F** = (600 lb)**i** − (300 lb)**j** − (200 lb)**k**, **M** = (3600 lb · ft)**i** + (9600 lb · ft)**j** − (3600 lb · ft)**k**. (*b*) **F** = (600 lb)**i** − (300 lb)**j** − (200 lb)**k**, **M** = −(1200 lb · ft)**i** − (3600 lb · ft)**k**.
3.62 (*a*) **F** = (600 lb)**i** − (300 lb)**j** + (600 lb)**k**, **M** = (3600 lb · ft)**i** − (3600 lb · ft)**k**. (*b*) **F** = (600 lb)**i** − (300 lb)**j** + (600 lb)**k**, **M** = −(1200 lb · ft)**i** − (9600 lb · ft)**j** − (3600 lb · ft)**k**.
3.63 Force-couple systems at *B* and *F*.
3.64 *c* and *f*.
3.65 Loading *e*.
3.66 (*a*) 0.6 m. (*b*) 1 m. (*c*) 1.8 m.
3.67 $x = QL/(P + Q)$. (*a*) x = 150 mm. (*b*) x = 300 mm.
3.68 12 kips ↓; 17.33 ft to the right of *A*.
3.69 1300 lb ↓; 8.69 ft to the right of *A*.
3.70 447 N ∠26.6°. (*a*) 100 mm to the left of *B*. (*b*) 50 mm above *B*.
3.71 (*a*) **R** = 3.79 kN→; $\mathbf{M}_D^R$ = 22.8 kN · m↺. (*b*) **R** = 3.79 kN→, 6 m below *DE*.
3.72 (*a*) **R** = 1878 lb ∠84.7°; $\mathbf{M}_A^R$ = 2450 lb · ft↺. (*b*) **R** = 1878 lb ∠84.7°, 14.0 ft above *A*.
3.73 100 lb ∖36.9° (*a*) At *A*. (*b*) 8 in. to the right of *B*. (*c*) 3 in. below *C*.
3.74 400 N ∖36.9° (*a*) 75 mm to the left of *C*. (*b*) 56.3 mm below *C*.
3.75 **R** = 0; **M** = 679 N · m↺.
3.76 **R** = −(100 lb)**i** − (900 lb)**j** − (200 lb)**k**; $\mathbf{M}_O^R$ = −(1200 lb · ft)**i** − (600 lb · ft)**k**.
3.77 (*a*) C_y = 15 lb, C_z = 0. (*b*) **R** = −(10 lb)**i** − (15 lb)**j**; **M** = −(180 lb · in.)**i**.
3.78 **R** = −(400 N)**j** − (200 N)**k**, **M** = (120 N · m)**i** + (40 N · m)**j** − (105 N · m)**k**. (*a*) Tightens. (*b*) Tightens.
3.79 **R** = −(400 N)**j** − (200 N)**k**, **M** = (160 N · m)**i** + (40 N · m)**j** − (105 N · m)**k**. (*a*) Tightens. (*b*) Loosens.
3.80 200 N; y = 63.4 mm, z = 200 mm.
3.81 72.2 N.
3.82 80 kips; x = 3.5 ft, z = 3 ft.
3.83 **A** = −(4 kips)**j**; **B** = −(28 kips)**j**.
3.84 **R** = −(75 lb)**j**; pitch = −0.667 in.; axis parallel to y axis at x = 5 in., z = 0.
3.85 (*a*) **R** = −(100 N)**i** + (50 N)**k**, **M** = −(25 N · m)**j** − (12.5 N · m)**k**. (*b*) **R** = 111.8 N, M_1 = −5.59 N · m; axis: $\theta_x = 153.4°$, $\theta_y = 90°$, $\theta_z = 63.4°$; axis intersects yz plane at y = −100 mm, z = 250 mm; pitch = −50 mm.
3.86 (*a*) $\mathbf{R} = \sqrt{2}P\mathbf{i}$; $\mathbf{M} = (Pa/\sqrt{2})(\mathbf{i} + \mathbf{j} - \mathbf{k})$. (*b*) $R = \sqrt{2}P$, $M_1 = Pa/\sqrt{2}$; parallel to the x axis at $y = z = \frac{1}{2}a$; pitch = $\frac{1}{2}a$.
3.87 $\mathbf{F}_B$ = −(20 lb)**i** + (30 lb)**j** + (60 lb)**k**, $\mathbf{F}_{xz}$ = (80 lb)**i** − (40 lb)**k**, at x = 1.50 in., $y = z = 0$.

CHAPTER 4

4.1 (*a*) $P = Wr/l \cos \theta$. (*b*) 40 lb.
4.2 **A** = 200 lb ↓; **B** = 200 lb ↑.
4.3 (*a*) 1.672 kN ↑. (*b*) 4.01 kN ↑.
4.4 30 kN ≤ P ≤ 210 kN.
4.5 60 lb ≤ P ≤ 560 lb.
4.6 22.5 lb ≤ P ≤ 627 lb.
4.7 T = 29.9 kips; **A** = 33.0 kips ∠31.5°.
4.8 (*a*) $T = (W \cos \theta)/(2 \cos \frac{1}{2}\theta)$. (*b*) 11.74 lb.
4.9 (*a*) 183.5 N. (*b*) 527 N ∠73.4°.
4.10 **A** = 665 N ∠30°; **B** = 665 N ∠30°.

4.11 **A** = 346 N ∠60.6°; **B** = 196.2 N ⦦30°.
4.12 **A** = 521 N ∠50.2°; **B** = 333 N←.
4.13 (*a*) 125 lb↓. (*b*) 325 lb ∠22.6°.
4.14 600 lb.
4.15 (*a*) **B** = 920 N↙; **C** = 80 N↙; **D** = 600 N↑. (*b*) Rollers 1 and 3.
4.16 (*a*) 706 N. (*b*) **A** = 441 N↑; **B** = 353 N→.
4.17 0.211 m.
4.18 T = 120 lb; **A** = 750 lb←; **B** = 750 lb→.
4.19 (*a*) **A** = 60.0 lb↑; **B** = 136.1 lb→; **C** = 32.2 lb←. (*b*) **A** = 0; **B** = 120.0 lb←; **C** = 240 lb→.
4.20 (*a*) **A** = 120 lb↑; **B** = 480 lb→; **C** = 480 lb←. (*b*) **A** = 103.9 lb↑; **B** = 356 lb→; **C** = 296 lb←.
4.21 $-49.1° \le \alpha \le +49.1°$.
4.22 (*a*) **B** = $2P/\sqrt{3}$ ∠60°; **C** = $2P/\sqrt{3}$ ⦦60°; **D** = P↓. (*b*) **B** = $\frac{1}{2}P$ ∠60°; **C** = $\frac{3}{2}P$ ⦦60°; **D** = $\sqrt{3}P/2$↓.
4.23 (*a*) $\mathbf{A}_x$ = 394 N→, $\mathbf{A}_y$ = 1541 N↑, $\mathbf{M}_A$ = 1891 N · m↻. (*b*) T_{max} = 619 N; T_{min} = 219 N.
4.24 (*a*) $\mathbf{A}_x$ = 1200 lb→, $\mathbf{A}_y$ = 300 lb↑, $\mathbf{M}_A$ = 4400 lb · ft↺. (*b*) 1928 lb ≤ W ≤ 2580 lb.
4.25 (*a*) **A** = $M/2L$↑; **B** = $M/2L$↓. (*b*) **A** = M/L←; **B** = M/L→. (*c*) **B** = 0; $\mathbf{M}_B$ = M↻. (*d*) **A** = M/L↑; **B** = M/L←; **D** = $\sqrt{2}\,M/L$ ⦨45°.
4.26 (*a*) **A** = 5540 N ∠87.3°; **C** = 683 N ⦩67.4°. (*b*) **A** = 4900 N↑; $\mathbf{M}_A$ = 1890 N · m↺. (*c*) **A** = 6740 N ∠83.6°; $\mathbf{M}_A$ = 3510 N · m↻; **C** = 1950 N ⦩67.4°.
4.27 (*a*) **A** = 32.8 lb ⦩66.1°; **B** = 38.7 lb←. (*b*) **A** = 50.0 lb↓; **B** = 55.7 lb ⦦21.0°. (*c*) **B** = 60.0 lb ⦩30°, $\mathbf{M}_B$ = 400 lb · in.↺.
4.28 (*a*) **A** = 59.1 lb ⦦28.3°; **B** = 58 lb↓. (*b*) **A** = 33.3 lb←; **B** = 35.3 lb ⦩58.1°. (*c*) **A** = 60 lb ⦩30°; $\mathbf{M}_A$ = 464 lb · in.↻.
4.29 (*a*) 6.77 kips. (*b*) 1.734 kips→; 4.17 kips↑.
4.30 (*a*) 14.58 ft. (*b*) $\mathbf{B}_x$ = 2.05 kips→; $\mathbf{B}_y$ = 4.20 kips↑.
4.31 (*a*) 377 N. (*b*) 849 N ⦦39.7°.
4.32 (*a*) $\tan\theta = (W/2P) - \cot\beta$. (*b*) 22.9°.
4.37 **A** = 534 N ∠69.4°; **E** = 187.5 N←.
4.38 (*a*) 36.9°. (*b*) **A** = 400 N↑; **E** = 300 N←.
4.39 **A** = 170 lb ∠28.1°; **B** = 150 lb←.
4.40 10 in.
4.41 **A** = 7.07 lb→; **B** = 40.6 lb ⦦80.0°.
4.42 **A** = 50 lb ∠30°; T = 50 lb.
4.43 (*a*) 1500 N ⦩30°. (*b*) 593 N ⦦30°.
4.44 **A** = 1077 N ∠21.8°; **E** = 1077 N ⦦21.8°.
4.45 $\theta = \cos^{-1}(2a/L)^{1/3}$.
4.46 (*a*) 30°. (*b*) 40°. (*c*) 60°.
4.47 (*a*) 49.1°. (*b*) 1.323W.
4.48 $\sin^3\theta = 2a/L$.
4.49 P = 462 N; **A** = (123 N)**j**; **B** = (1139 N)**j**.
4.50 P = 400 N; **A** = (192 N)**j** − (120 N)**k**, **B** = (954 N)**j** + (320 N)**k**.
4.51 (*a*) T_B = 24 lb; T'_B = 12 lb. (*b*) A_y = 55.2 lb, A_z = −12.49 lb; E_y = 33.4 lb, E_z = −2.50 lb; A_x and E_x indeterminate ($A_x + E_x = 0$).
4.52 (*a*) T_B = 27 lb; T'_B = 9 lb. (*b*) A_y = 27.4 lb, A_z = −4.18 lb; E_y = 5.47 lb, E_z = 37.6 lb; A_x and E_x indeterminate ($A_x + E_x = 0$).
4.53 T_A = 24.5 N; T_B = 73.6 N; T_C = 98.1 N.
4.54 m = 25.9 kg; x = 397 mm, z = 324 mm.
4.55 **A** = $\frac{1}{2}wL$↑; **C** = $\frac{1}{6}wL$↑; **D** = $\frac{4}{3}wL$↑.
4.56 $\frac{1}{3}L$.
4.57 $T_{BD} = T_{BE}$ = 11 kN; **A** = −(3.6 kN)**i** + (14 kN)**j**.
4.58 T_{BC} = 750 N; T_{BD} = 150 N; **A** = 700 N along AB.
4.59 T_{BC} = 560 lb; T_{BD} = 360 lb; **A** = 900 lb ∠36.9° (along AB).
4.60 T_{CD} = 765 lb; T_{EBF} = 990 lb.
4.61 T = 37.5 lb; **A** = (36.3 lb)**i** + (65.6 lb)**j**; **B** = (75 lb)**j**.
4.62 T = 846 lb; A_x = 0, A_y = −83.3 lb; B_x = 400 lb, B_y = 250 lb; A_z and B_z indeterminate ($A_z + B_z$ = 667 lb).
4.63 F_{CD} = 19.62 N; **A** = −(19.23 N)**i** + (45.1 N)**j**; **B** = (49.1 N)**j**.
4.64 P = 118.9 N; **A** = (42.9 N)**i** − (69.9 N)**k**; **B** = (61.1 N)**i** + (196.2 N)**j** + (84.7 N)**k**.
4.65 T = 37.5 lb; **B** = (36.3 lb)**i** + (140.6 lb)**j**, $\mathbf{M}_B$ = −(196.9 lb · ft)**i** + (108.9 lb · ft)**j**.

4.66 $T = 846$ lb;
$\mathbf{B} = (400 \text{ lb})\mathbf{i} + (166.7 \text{ lb})\mathbf{j} + (667 \text{ lb})\mathbf{k}$;
$\mathbf{M}_B = (833 \text{ lb} \cdot \text{ft})\mathbf{i}$.

4.67 $F_{CD} = 19.62$ N;
$\mathbf{B} = -(19.22 \text{ N})\mathbf{i} + (94.2 \text{ N})\mathbf{j}$;
$\mathbf{M}_B = -(40.6 \text{ N} \cdot \text{m})\mathbf{i} - (17.30 \text{ N} \cdot \text{m})\mathbf{j}$.

4.68 $\mathbf{A} = (600 \text{ N})\mathbf{j} - (750 \text{ N})\mathbf{k}$;
$\mathbf{B} = (900 \text{ N})\mathbf{i} + (750 \text{ N})\mathbf{k}$;
$\mathbf{C} = -(900 \text{ N})\mathbf{i} + (600 \text{ N})\mathbf{j}$.

4.69 $\mathbf{B} = (60 \text{ N})\mathbf{k}$; $\mathbf{C} = (30 \text{ N})\mathbf{j} - (16 \text{ N})\mathbf{k}$;
$\mathbf{D} = -(30 \text{ N})\mathbf{j} + (4 \text{ N})\mathbf{k}$.

4.70 $\mathbf{B} = (60 \text{ N})\mathbf{k}$; $\mathbf{C} = -(12 \text{ N})\mathbf{k}$,
$\mathbf{M}_C = (12 \text{ N} \cdot \text{m})\mathbf{j} + (90 \text{ N} \cdot \text{m})\mathbf{k}$.

4.71 $T_{BD} = 1560$ lb; $T_{BE} = T_{CF} = 1300$ lb;
$\mathbf{A} = (3840 \text{ lb})\mathbf{i} - (600 \text{ lb})\mathbf{k}$.

4.72 $T_{BE} = 1300$ lb; $T_{CE} = 1432$ lb; $T_{CF} = 0$;
$\mathbf{A} = (2400 \text{ lb})\mathbf{i} - (600 \text{ lb})\mathbf{k}$.

4.73 420 lb.

4.74 206 N.

4.75 231 N.

4.76 36 lb.

CHAPTER 5

5.1 $\bar{x} = 55.4$ mm, $\bar{y} = 93.8$ mm.

5.2 $\bar{x} = 55.4$ mm, $\bar{y} = 66.2$ mm.

5.3 $\bar{x} = 1.333$ in., $\bar{y} = 2.83$ in.

5.4 $\bar{x} = 3.55$ in., $\bar{y} = 4$ in.

5.5 $\bar{x} = 0$, $\bar{y} = 9.26$ mm.

5.6 $\bar{x} = 0$, $\bar{y} = -20.2$ mm.

5.7 $\bar{x} = 1.613$ in., $\bar{y} = 0$.

5.8 $\bar{x} = 3.18$ in., $\bar{y} = 6$ in.

5.9 $\bar{x} = 6$ in., $\bar{y} = 4.80$ in.

5.10 $\bar{x} = 0$, $\bar{y} = 4.57$ ft.

5.11 $\bar{x} = 152.5$ mm, $\bar{y} = 17.2$ mm.

5.12 $\bar{x} = 0$, $\bar{y} = -149.3$ mm.

5.13 $\bar{x} = 53.0$ mm, $\bar{y} = 91.5$ mm.

5.14 $\bar{x} = 53.0$ mm, $\bar{y} = 68.6$ mm.

5.15 $\bar{x} = 1.468$ in., $\bar{y} = 0$.

5.16 $\bar{x} = 3.19$ in., $\bar{y} = 6$ in.

5.17 26.6°.

5.18 21.4 in.

5.19 27.3 in.

5.20 (*a*) 12.0°. (*b*) 39.3°.

5.21 $\bar{x} = 0.56$ in., $\bar{y} = 0$.

5.22 $\bar{x} = 1.13$ in., $\bar{y} = 0$.

5.23 $\bar{x} = 0$, $\bar{y} = 59$ mm.

5.24 $\bar{x} = 168$ mm.

5.25 $\bar{x} = \frac{1}{2}a$, $\bar{y} = \frac{2}{5}h$.

5.26 $\bar{x} = \bar{y} = \frac{9}{20}a$.

5.27 $\bar{x} = \frac{2}{5}a$, $\bar{y} = \frac{1}{2}b$.

5.28 $\bar{x} = 0.611L$.

5.29 $\bar{x} = 0.300a$.

5.30 $\bar{y} = 0.310h$.

5.31 $\bar{x} = 1.820a$, $\bar{y} = 0.303a$.

5.32 $\bar{y} = \dfrac{\pi - 2}{\pi}L$, $\bar{y} = \dfrac{\pi}{8}a$.

5.33 (*a*) $\frac{1}{2}\pi a^2 h$. (*b*) $\frac{8}{15}\pi a h^2$.

5.34 $A = 2\pi^2 Rr + 2\pi r^2$, $V = \pi^2 R r^2$.

5.35 $4\pi R^2 \sin^2 (\phi/2)$.

5.36 (*a*) 6.48×10^6 mm^3. (*b*) 5.43×10^6 mm^3.

5.37 0.655 m^3.

5.38 1998 ft^3.

5.39 (*a*) 867 ft^3. (*b*) 1093 ft^3.

5.40 300×10^3 mm^3.

5.41 151.5 mm.

5.42 0.1089 in^3.

5.43 (*a*) 26.3 in^3. (*b*) 57.9 in^2.

5.44 $V = 248 \times 10^3$ mm^3;
$A = 34.2 \times 10^3$ mm^2.

5.45 $\mathbf{R} = 9.45$ kN$\downarrow$, 2.57 m to the right of A;
$\mathbf{A} = 4.05$ kN$\uparrow$, $\mathbf{B} = 5.40$ kN$\uparrow$.

5.46 $\mathbf{R} = 32$ kN$\downarrow$, 3 m to the right of A;
$\mathbf{A} = 20$ kN$\uparrow$, $\mathbf{B} = 12$ kN$\uparrow$.

5.47 $\mathbf{B} = 550$ lb$\uparrow$; $\mathbf{M}_B = 2300$ lb · in.$\circlearrowright$.

5.48 $\mathbf{A} = 10{,}800$ lb$\uparrow$, $\mathbf{B} = 3{,}600$ lb$\uparrow$.

5.49 $\mathbf{B} = 1200$ N$\uparrow$; $\mathbf{M}_B = 800$ N · m$\circlearrowleft$.

5.50 $\mathbf{A} = 2.5$ kN$\uparrow$, $\mathbf{B} = 2$ kN$\uparrow$.

5.51 (*a*) 0.500; $\mathbf{A} = 0$, $\mathbf{M}_A = \frac{1}{4}w_A L^2 \circlearrowright$.
(*b*) 1.000; $\mathbf{A} = \frac{1}{3}w_A L\uparrow$, $\mathbf{M}_A = 0$.

5.52 (*a*) $w_2 = \dfrac{w_1 a}{L}\left(4 - 3\dfrac{a}{L}\right)$;
$w_3 = \dfrac{w_1 a}{L}\left(\dfrac{3a}{L} - 2\right)$. (*b*) $a/L \geq \frac{2}{3}$.

5.53 (*a*) $\mathbf{H} = 17{,}970$ lb$\rightarrow$, $\mathbf{V} = 60{,}100$ lb$\uparrow$;
12.63 ft to the right of A.
(*b*) $\mathbf{R} = 19{,}190$ lb $\measuredangle$20.6°.

5.54 0.450 m.

5.55 90 mm.

5.56 (*a*) $\mathbf{R} = 1966$ lb←, 1.714 ft below *A*.
(*b*) 1123 lb→.
5.57 $A_x = 1.635$ kN←, $A_y = 9.81$ kN↑; $\mathbf{D} = 8.18$ kN←.
5.58 $\tan\theta = \frac{2}{3}\gamma b r^2/W$.
5.59 $\mathbf{A} = 228$ lb↑; $\mathbf{B}_x = 211$ lb←, $\mathbf{B}_y = 333$ lb↑.
5.60 1.732.
5.61 1.491 m.
5.62 3.56 m.
5.63 2.94.
5.64 0.792.
5.65 $\sqrt{3}$.
5.66 0.707.
5.67 $\frac{21}{16}h$ above tip of cone.
5.68 69.5 mm above base.
5.69 $\bar{x} = 0.610$ in., $\bar{y} = -2.34$ in., $\bar{z} = 0$.
5.70 $\bar{x} = \bar{z} = 0.1803$ in., $\bar{y} = 0$.
5.71 $\bar{x} = 1.875$ mm, $\bar{y} = -7.50$ mm, $\bar{z} = 0$.
5.72 $\bar{x} = 168.5$ mm, $\bar{y} = -53.3$ mm, $\bar{z} = 29.6$ mm.
5.73 $\bar{x} = 4$ in., $\bar{y} = -1.725$ in., $\bar{z} = 0.589$ in.
5.74 $\bar{x} = 0.0729$ in., $\bar{y} = -1.573$ in., $\bar{z} = 0$.
5.75 $\bar{x} = 93.9$ mm, $\bar{y} = 3.40$ mm, $\bar{z} = 0$.
5.76 $\bar{x} = \bar{z} = 129.7$ mm, $\bar{y} = 174.0$ mm.
5.77 0.706 ft.
5.78 1.870 in. below top.
5.79 1.457 in. above base.
5.80 82.2 mm above the base.
5.81 $\bar{x} = \dfrac{2n+1}{2n+2}h$, $\bar{y} = \bar{z} = 0$.
5.82 $\bar{x} = \frac{5}{6}h$, $\bar{y} = 20a/21\pi$, $\bar{z} = 0$.
5.83 $\bar{x} = \bar{z} = 0$, $\bar{y} = -\dfrac{a^2 - ah + \frac{1}{4}h^2}{a - \frac{1}{3}h}$.
5.84 0.125 m below center of tank.
5.85 $\bar{x} = \frac{2}{3}a$, $\bar{y} = \frac{2}{3}b$, $\bar{z} = \frac{2}{3}h$.
5.86 $\bar{x} = \frac{1}{2}a$, $\bar{y} = \pi^2 h/32$, $\bar{z} = \frac{1}{2}b$.

CHAPTER 6

6.1 $F_{AB} = 1733$ lb T; $F_{BC} = 2133$ lb T; $F_{CA} = 2667$ lb C.
6.2 $F_{AB} = 1600$ lb C; $F_{AC} = 2000$ lb T; $F_{BC} = 1709$ lb T.
6.3 $F_{AB} = 4.80$ kN T; $F_{AC} = 5.20$ kN T; $F_{AD} = 4$ kN C; $F_{BD} = 5.20$ kN C; $F_{CD} = 4.80$ kN C.
6.4 $F_{AB} = 3900$ N T; $F_{AC} = 4500$ N C; $F_{BC} = 3600$ N C.
6.5 $F_{AB} = F_{DE} = F_{BG} = F_{DI} = 0$; $F_{AF} = F_{CH} = F_{EJ} = 400$ N C; $F_{BC} = F_{CD} = 800$ N C; $F_{BF} = F_{DJ} = 849$ N C; $F_{BH} = F_{DH} = 283$ N T; $F_{FG} = F_{GH} = F_{HI} = F_{IJ} = 600$ N T.
6.6 $F_{AB} = 1800$ lb T; $F_{AC} = 0$; $F_{BC} = 2250$ lb C; $F_{BD} = 1350$ lb T; $F_{CD} = 3600$ lb T; $F_{CE} = 1350$ lb C; $F_{DE} = 4500$ lb C; $F_{DF} = 4050$ lb T.
6.7 $F_{AB} = 0$; $F_{AC} = 22.4$ kips T; $F_{AD} = 41.2$ kips T; $F_{BC} = 60$ kips C; $F_{CD} = 40$ kips C.
6.8 $F_{AB} = F_{BC} = 2$ kN C; $F_{AD} = F_{CF} = 4$ kN C; $F_{AE} = F_{CE} = 2.5$ kN T; $F_{BE} = 1.5$ kN C; $F_{DE} = F_{EF} = 0$.
6.9 $F_{AB} = F_{AF} = F_{FG} = 5$ kN C; $F_{AD} = F_{BD} = F_{DF} = F_{DG} = 7.07$ kN T; $F_{BC} = F_{GH} = 5$ kN T; $F_{BE} = F_{CE} = F_{EG} = F_{EH} = 7.07$ kN C.
6.10 $F_{AC} = 390$ lb T; $F_{BC} = 0$; $F_{BD} = 360$ lb C; $F_{CD} = 150$ lb T; $F_{CE} = 360$ lb T; $F_{DE} = 390$ lb C.
6.11 *1-2* = *2-4* = *4-6* = *6-8* = 30 kips T; *1-3* = *3-5* = *5-7* = 34 kips C; *2-3* = *3-4* = *4-5* = *5-6* = 0; *6-7* = 24 kips C; *7-8* = 50 kips C.
6.12 $F_{AB} = F_{DE} = 8$ kN C; $F_{AF} = F_{FG} = F_{GH} = F_{HE} = 6.93$ kN T; $F_{BC} = F_{CD} = F_{BG} = F_{DG} = 4$ kN C; $F_{BF} = F_{DH} = F_{CG} = 4$ kN T.
6.13 All simple trusses except 6.14 and 6.15.
6.14 *BC, CD, IJ, IL, KL, LM, MN*.
6.15 *AF, CH, DE, EJ, GL, IN*.
6.16 *AG, AH, BD, DH, BE, EJ, FJ*.
6.17 $F_{BD} = 36$ kips C; $F_{CD} = 45$ kips C.
6.18 $F_{DF} = 60$ kips C; $F_{DG} = 15$ kips C.
6.19 $F_{DF} = 91.4$ kN T; $F_{DE} = 38.6$ kN C.
6.20 $F_{CD} = 64.3$ kN T; $F_{CE} = 92.1$ kN C.
6.21 $F_{FH} = 16.97$ kips T; $F_{GH} = 12$ kips C; $F_{GI} = 18$ kips C.

6.22 $F_{DF} = 8.25$ kips T; $F_{DE} = 3$ kips C; $F_{CE} = 8$ kips C.

6.23 $F_{BD} = 15$ kN C; $F_{CD} = 35$ kN C; $F_{CE} = 15$ kN T.

6.24 $F_{EG} = 45$ kN T; $F_{EF} = 55$ kN C; $F_{DF} = 45$ kN C.

6.25 $F_{EF} = 2400$ lb T; $F_{FG} = 1500$ lb C; $F_{GI} = 2600$ lb C.

6.26 $F_{EC} = 4690$ lb T; $F_{CD} = 3600$ lb C; $F_{CB} = 0$.

6.27 $F_{BD} = 37.5$ kN T; $F_{DE} = 22.5$ kN T.

6.28 $F_{FH} = 12.5$ kN T; $F_{DH} = 90$ kN T.

6.29 $F_{AB} = 5P/6$ *ten.*, $F_{KL} = 7P/6$ *ten.*

6.30 22.5 kN C.

6.31 $F_{AB} = 0$; $F_{EJ} = \frac{2}{3}P$ *comp.*

6.32 $F_{AB} = 0$; $F_{EJ} = \frac{2}{3}P$ *comp.*

6.33 $F_{BD} = 64$ kN C; $F_{CE} = 48$ kN T; $F_{BE} = 20$ kN T.

6.34 $F_{BD} = 21.3$ kN C; $F_{CE} = 10.67$ kN T. $F_{CD} = 13.33$ kN T.

6.35 $F_{CE} = 15$ kips T; $F_{CG} = 19$ kips C; $F_{BE} = 10$ kips C.

6.36 $F_{BE} = 16$ kips C; $F_{CG} = 10$ kips C; $F_{BG} = 10$ kips T.

6.37 $F_{BD} = 1750$ N C; $\mathbf{C}_x = 1400$ N$\leftarrow$, $\mathbf{C}_y = 700$ N$\downarrow$.

6.38 $F_{BD} = 300$ lb T; $\mathbf{C}_x = 150$ lb$\leftarrow$, $\mathbf{C}_y = 180$ lb$\uparrow$.

6.39 (*a*) $\mathbf{B}_x = 250$ lb$\rightarrow$, $\mathbf{B}_y = 100$ lb$\downarrow$. (*b*) $\mathbf{B} = 100$ lb$\uparrow$.

6.40 (*a*) $\mathbf{A} = 130$ N $\measuredangle$22.6°, $\mathbf{B} = 130$ N $\measuredangle$22.6°. (*b*) $\mathbf{A} = 60$ N$\rightarrow$, $\mathbf{B} = 60$ N$\leftarrow$. (*c*) $\mathbf{A} = 39.1$ N $\measuredangle$39.8°, $\mathbf{B} = 128.5$ N $\measuredangle$76.5°.

6.41 $\mathbf{A}_x = 250$ lb$\leftarrow$, $\mathbf{A}_y = 600$ lb$\uparrow$; $\mathbf{C}_x = 250$ lb$\rightarrow$, $\mathbf{C}_y = 600$ lb$\uparrow$; $\mathbf{B} = 790$ lb$\leftarrow$.

6.42 $\mathbf{E} = 80$ lb$\downarrow$; $\mathbf{F}_x = 90$ lb$\rightarrow$, $\mathbf{F}_y = 120$ lb$\uparrow$; $\mathbf{G}_x = 90$ lb$\leftarrow$, $\mathbf{G}_y = 40$ lb$\downarrow$.

6.43 (*a*) *At each wheel:* $\mathbf{A} = 117.5$ kN$\uparrow$; $\mathbf{B} = 176.9$ kN$\uparrow$. (*b*) $\mathbf{C} = 8.28$ kN$\rightarrow$; $\mathbf{D}_x = 8.28$ kN$\leftarrow$, $\mathbf{D}_y = 256$ kN$\downarrow$.

6.44 $\mathbf{A}_x = 4.5$ kN$\leftarrow$, $\mathbf{A}_y = 2.5$ kN$\downarrow$; $\mathbf{F}_{CF} = 3.75$ kN$\rightarrow$; $\mathbf{E}_x = 3.75$ kN$\leftarrow$; $\mathbf{E}_y = 2.5$ kN$\uparrow$.

6.45 (*a*) $\mathbf{C}_x = 16$ kN$\leftarrow$, $\mathbf{C}_y = 14$ kN$\uparrow$. (*b*) $\mathbf{B}_x = 16$ kN$\leftarrow$, $\mathbf{B}_y = 4$ kN$\uparrow$.

6.46 (*a*) $\mathbf{C}_x = 12$ kN$\leftarrow$, $\mathbf{C}_y = 6$ kN$\uparrow$. (*b*) $\mathbf{B}_x = 12$ kN$\leftarrow$, $\mathbf{B}_y = 6$ kN$\uparrow$.

6.47 $\mathbf{A} = 480$ lb$\rightarrow$; $\mathbf{B} = 80$ lb$\leftarrow$; $\mathbf{C} = 500$ lb $\measuredangle$36.9°.

6.48 $\mathbf{F} = 495$ lb $\measuredangle$76.0°; $F_{AE} = 400$ lb C; $F_{BD} = 200$ lb T.

6.49 $\mathbf{C} = 632$ N $\measuredangle$71.6°; $\mathbf{D} = 250$ N $\measuredangle$36.9°; $F_{AE} = 750$ N C; $F_{BF} = 500$ N T.

6.50 $\mathbf{C} = 750$ N $\measuredangle$36.9°; $\mathbf{D} = 671$ N $\measuredangle$26.6°; $F_{AE} = 1500$ N C; $F_{BF} = 2250$ N T.

6.51 $\mathbf{B} = 17.07$ lb$\leftarrow$; $\mathbf{E}_x = 36.3$ lb$\rightarrow$, $\mathbf{E}_y = 32.0$ lb$\downarrow$; $\mathbf{H}_x = 19.20$ lb$\leftarrow$, $\mathbf{H}_y = 32.0$ lb$\uparrow$.

6.52 $F_{AB} = F_{BC} = 1600$ lb T; $F_{AE} = 2000$ lb T; $F_{BF} = 0$; $F_{DE} = 3200$ lb C; $F_{CF} = 1200$ lb C; $F_{CG} = 2000$ lb T.

6.53 $\mathbf{C}_x = 2.5$ kN$\rightarrow$, $\mathbf{C}_y = 1$ kN$\uparrow$; $\mathbf{D} = 2.5$ kN$\leftarrow$; $\mathbf{E} = 2.5$ kN$\leftarrow$; $\mathbf{F}_x = 2.5$ kN$\rightarrow$, $\mathbf{F}_y = 1$ kN$\uparrow$.

6.54 $\mathbf{C}_x = 1800$ N$\leftarrow$, $\mathbf{C}_y = 1350$ N$\uparrow$; $\mathbf{D}_x = 1800$ N$\rightarrow$, $\mathbf{D}_y = 2025$ N$\downarrow$; $\mathbf{E} = 675$ N$\uparrow$; $\mathbf{F} = 0$.

6.55 $\mathbf{E} = 528$ lb $\measuredangle$16.5°; $\mathbf{F} = 677$ lb $\measuredangle$41.6°; $\mathbf{C} = 1356$ lb $\measuredangle$41.6°.

6.56 *On the frame:* $\mathbf{F} = 1611$ lb$\uparrow$; $\mathbf{A} = 500$ lb$\leftarrow$; $\mathbf{D}_x = 500$ lb$\rightarrow$, $\mathbf{D}_y = 861$ lb$\downarrow$.

6.57 (*a*) 1200 N$\rightarrow$. (*b*) 1230 N $\measuredangle$12.7°.

6.58 (*a*) 2.5 kN. (*b*) 2.76 kN $\measuredangle$63.1°.

6.59 $\mathbf{C} = 150$ lb$\rightarrow$; $\mathbf{D} = 250$ lb $\measuredangle$53.1°.

6.60 14,800 lb.

6.61 (*a*) 128.0 N. (*b*) 510 N.

6.62 45.4 N · m$\circlearrowleft$.

6.63 18.75 lb.

6.64 $F_{AB} = 18.97$ kips C; $F_{CD} = 4.27$ kips T; $F_{EF} = 9.61$ kips C.

6.65 (*a*) $-(6\text{ N}\cdot\text{m})\mathbf{i}$. (*b*) $\mathbf{G} = 0$, $\mathbf{M}_G = -(5.40\text{ N}\cdot\text{m})\mathbf{i}$; $\mathbf{H} = 0$, $\mathbf{M}_H = -(12.60\text{ N}\cdot\text{m})\mathbf{i}$.

6.66 (*a*) $-(21\text{ N}\cdot\text{m})\mathbf{i}$. (*b*) $\mathbf{G} = 0$, $\mathbf{M}_G = (2.70\text{ N}\cdot\text{m})\mathbf{i}$; $\mathbf{H} = 0$, $\mathbf{M}_H = -(11.70\text{ N}\cdot\text{m})\mathbf{i}$.

6.67 $T_1 = 350$ lb; $T_2 = 4150$ lb.

6.68 18.4°.
6.69 88.3 kN *C*.
6.70 68.9 kN *C*.
6.71 (*a*) 140 lb · ft↻. (*b*) 60 lb · ft↻.
6.72 (*a*) 600 lb ↓. (*b*) 1400 lb ↓.
6.73 50 lb.
6.74 (*a*) zero. (*b*) $\mathbf{C}_x$ = 4500 lb→, $\mathbf{C}_y$ = 1500 lb ↓.
6.75 (*a*) 57.7 N · m. (*b*) **B** = 0; **D** = −(144.0 N)**k**; **E** = (144.0 N)**k**.
6.76 (*a*) M_A = 43.3 N · m. (*b*) **B** = −(250 N)**k**; **D** = (350 N)**k**; **E** = −(100 N)**k**.

CHAPTER 7

7.1 (On *AJ*) **F** = 397 lb↙; **V** = 514 lb↘; **M** = 2370 lb · ft↺.
7.2 (On *JG*) **F** = 90 lb→; **V** = 40 lb ↑; **M** = 60 lb · ft↺.
7.3 (On *AJ*) **F** = 21.5 kN ∠26.6°; **V** = 7.16 kN ∠63.4°; **M** = 10 kN · m↺.
7.4 (On *AJ*) **F** = 16.99 kN ∠26.6°; **V** = 7.15 kN ∠63.4°; **M** = 12.5 kN · m↺.
7.5 (On left-hand portion) **F** = 300 N←; **V** = 50 N ↓; **M** = 40 N · m↺.
7.6 225 mm.
7.7 (On *CB*) (*a*) **F** = $\frac{1}{2}wL \csc \theta$↗; **V** = 0; **M** = $\frac{1}{8}wL^2 \cos \theta$↻. (*b*) **F** = 250 lb↗; **V** = 0; **M** = 450 lb · ft↻.
7.8 (On *AC*) (*a*) **F** = **V** = 0; **M** = $\frac{1}{8}wL^2 \cos \theta$↺. (*b*) **F** = **V** = 0; **M** = 450 lb · ft↺.
7.9 (On left section) **F** = 206 lb↙; **V** = 15.81 lb↖; **M** = 300 lb · in.↻.
7.10 (*a*) **F** = T←; **V** = $\frac{1}{2}P$ ↓; **M** = $\frac{1}{4}PL - Th$↺. (*b*) *PL*/4*h*.
7.11 (On *AJ*) **F** = 0; **V** = 250 N→; **M** = 25 N · m↻.
7.12 (On *AJ*) **F** = 500 N ↓; **V** = 187.5 N←; **M** = 18.75 N · m↺.
7.13 (On *JB*) **F** = 0; **V** = 90 lb ↑; **M** = 900 lb · in.↺.
7.14 (On *KD*) **F** = 0; **V** = 30 lb ↓; **M** = 300 lb · in.↺.
7.15 (*a*) $M_{max} = PL/4$, at $a = L/2$. (*b*) $\frac{1}{4}L/\sqrt{1 + (L/h)^2}$.
7.16 (On *JB*) (*a*) $\mathbf{M}_J = \frac{1}{2}Wr$↻. (*b*) $\mathbf{M}_{max}$ = 0.579 *Wr*↻, θ = 116.3°.
7.17 $M_A = -\frac{1}{2}wL^2$.
7.18 $M = +T$ between *B* and *C*; $M = 0$ everywhere else.
7.19 $M_{max} = w(\frac{1}{8}L^2 - \frac{1}{2}a^2)$.
7.20 $M = \frac{1}{2}wa^2$ between *B* and *C*.
7.21 M_{max} = +25.2 kN · m.
7.22 M_C = +800 lb · ft.
7.23 M_C = −3000 lb · ft; M_D = +1200 lb · ft.
7.24 M_D = +28 kN · m.
7.25 3.43 ft.
7.26 24 kN.
7.27 M_D = +144 lb · in.
7.28 Just to the left of *D*: *M* = +20 N · m.
7.29 Just to the right of *D*: *M* = +180 N · m.
7.30 Just to the left of *D*: *M* = +120 lb · ft.
7.31 M_C = −4 kN · m.
7.32 M_C = −3500 lb · ft.
7.41 M_D = +60 kip · ft.
7.42 M_D = +2.20 kN · m.
7.43 M_D = −12 N · m.
7.44 M_D = +530 lb · in.
7.45 $M_{max} = +\frac{1}{2}wa^2\left(1 - \frac{a}{2L}\right)^2$, at $x = a\left(1 - \frac{a}{2L}\right)$.
7.46 *M* = −2 kN · m at *C*; *M* = +0.667 kN · m, 1.333 m to the left of *B*.
7.47 *M* = −10 kN · m at *A*; *M* = +3.5 kN · m, 3 m to the right of *A*.
7.48 *M* = +2000 lb · ft, 10 ft to the right of *A*.
7.49 (*a*) $V = -c$ and $M = 0$ everywhere. (*b*) $V = 0$ everywhere; $M_A = -cL$, $M_B = 0$.
7.50 V_D = +1 kN; M_C = −0.4 kN · m.
7.51 $M_C = M_D = M_E = M_F$ = −12.5 kip · ft.
7.52 8 in.
7.53 (*a*) $\mathbf{E}_x$ = 1500 lb→, $\mathbf{E}_y$ = 1000 lb ↑. (*b*) 1803 lb.
7.54 7.50 ft.
7.55 $\mathbf{E}_x$ = 1600 N→, $\mathbf{E}_y$ = 1100 N ↑.

7.56 1 m.
7.57 3.15 ft.
7.58 671 N.
7.59 (*a*) 395 N. (*b*) 1456 N.
7.60 22.5 ft.
7.61 $T_{max} = 34.4$ kN; $T_{min} = 28.0$ kN.
7.62 $h = 0.1429$ ft; $\theta_A = 38.2°$, $\theta_C = 35.5°$.
7.63 $h = 0.273$ ft; $\theta_A = 39.3°$, $\theta_C = 34.3°$.
7.64 8 m to left of *B*; 14.72 kN.
7.65 164.8 m; 6130 N.
7.66 26.7 m; 345 N.
7.67 (*a*) 345 ft. (*b*) 212.5 lb.
7.68 (*a*) 309.2 ft. (*b*) 32.6 ft. (*c*) 191.3 lb.
7.69 3.98 m; 100.7 N.
7.70 (*a*) 40.6 N. (*b*) 2.68 m. (*c*) 106.1 N.
7.71 $h = 7.96$ ft, $T_m = 20.5$ lb.
7.72 (*a*) $1.029(2\pi r)$. (*b*) $0.657wr$.

CHAPTER 8

8.1 74.5 lb.
8.2 5.77 lb.
8.3 $P = W \sin(\alpha + \phi_s)$, $\theta = \alpha + \phi_s$.
8.4 (*a*) 116.2 N ∠36.3°. (*b*) 46.5 N ∠13.7°.
8.5 (*a*) 3 lb. (*b*) 4.5 lb. (*c*) 6 lb.
8.6 (*a*) *A* and *B* move.
(*b*) $\mathbf{F}_A = 8.53$ N↖; $\mathbf{F}_B = 9.48$ N↖.
8.7 137.2 N.
8.8 $\theta \geq 29.4°$.
8.9 51.4 lb←.
8.10 40 lb→.
8.11 (*a*) 2340 N. (*b*) 3060 N.
8.12 (*a*) 54.5 N · m. (*b*) 46.2 N · m.
8.13 0.175.
8.14 0.80.
8.15 $\mu_A = 0.222$; $\mu_C = 0.247$.
8.16 0.353.
8.17 834 N.
8.18 371 N (motion impends at *C*).
8.19 0.955 lb.
8.20 12.21 lb.
8.21 1/7.
8.22 0.1940*L*.
8.23 (*a*) 8 lb. (*b*) 12 lb.
8.24 2.80 m.
8.25 (*a*) $P = W \tan \phi / \cos \theta$.
(*b*) $M = \frac{1}{2} Wr \sin 2\phi / \cos \theta$.
8.26 544 lb · in.↰.
8.27 376 lb · in.↰.
8.28 0.520 lb↗; *A* rolls.
8.29 1.300 lb↗; *C* rolls.
8.30 (*a*) 295 lb←.
(*b*) $\mathbf{B}_x = 305$ lb←, $\mathbf{B}_y = 500$ lb↓.
8.31 (*a*) 111.3 lb→.
(*b*) $\mathbf{B}_x = 511$ lb←, $\mathbf{B}_y = 500$ lb↓.
8.32 1.858 kN.
8.33 1.935 kN↓.
8.34 190.4 N.
8.35 (*a*) 75 lb→. (*b*) Machine will move.
8.36 (*a*) 58.5 lb←. (*b*) Machine will not move.
8.37 (*a*) Wedge will be forced up and out.
(*b*) Wedge will self-lock.
8.38 16.86 N · m.
8.39 7.13 N · m.
8.40 56.4 lb · in.
8.41 42.0 lb · in.
8.42 (*a*) 0.24. (*b*) 55.6 lb.
8.43 (*a*) 235 N. (*b*) 207 N.
8.44 0.10.
8.45 4.69 lb/ton.
8.46 120 mm.
8.47 0.160.
8.48 3.47 lb.
8.49 195 N · m.
8.50 577 N.
8.51 58.9 N.
8.52 0.083 in.
8.53 7.19 lb/ton.
8.54 300 mm.
8.55 (*a*) 0.293. (*b*) 200 lb.
8.56 (*a*) 0.33. (*b*) 2.66 turns.
8.57 (*b*) 952 mm.
8.58 3.96 m.
8.59 0.116.
8.60 69.4 N · m.
8.61 27.7 N · m.
8.62 88.8 N · m↰.
8.63 0.35.
8.64 61.0 lb · ft.
8.65 71.5 lb · ft.
8.66 412 lb.

CHAPTER 9

9.1 $2a^3b/7$.
9.2 $a^3b/28$.
9.3 $a^3b/20$.
9.4 $2ab^3/15$.
9.5 $ab^3/20$.
9.6 $ab^3/28$.
9.7 $ab^3/30$; $b\sqrt{2/15}$.
9.8 $\pi ab^3/8$; $\frac{1}{2}b$.
9.9 $ab^3/9$; $b\sqrt{5/27}$.
9.10 $a^3b/6$; $a\sqrt{2/3}$.
9.11 $\pi a^3b/8$; $\frac{1}{2}a$.
9.12 $3a^3b/11$; $a\sqrt{5/11}$.
9.13 $a\sqrt{2}/4$.
9.14 $(b^2 + h^2)bh/3$, $\sqrt{(b^2 + h^2)/3}$.
9.15 $5a^4/12$; $a\sqrt{5/12}$.
9.16 $\pi ab(a^2 + b^2)/8$, $\frac{1}{2}\sqrt{a^2 + b^2}$.
9.17 $4a^3/9$.
9.18 $0.935a$.
9.19 $xy^2 =$ constant.
9.20 501×10^6 mm^4; 149.4 mm.
9.21 150.3×10^6 mm^4; 81.9 mm.
9.22 382 in^4; 4.04 in.
9.23 185.4 in^4, 2.81 in.
9.24 (*a*) $I_x = 174.7$ in^4, $I_y = 1851$ in^4.
(*b*) 22.4 in.
9.25 (*a*) 10.08×10^6 mm^4.
(*b*) 29.2×10^6 mm^4.
(*c*) 1.235×10^6 mm^4.
9.26 16×10^3 mm^2; 160×10^6 mm^4.
9.27 565×10^6 mm^4.
9.28 (*a*) 159.5×10^6 mm^4.
(*b*) 31.8×10^6 mm^4.
9.29 (*a*) 185.9 in^4. (*b*) 154.0 in^4.
9.30 379 in^4.
9.31 $\bar{I}_x = 292.5$ in^4, $\bar{I}_y = 86.8$ in^4.
9.32 $\bar{I}_x = \bar{I}_y = 2.08 \times 10^6$ mm^4.
9.33 $\bar{I}_x = 2.04 \times 10^6$ mm^4,
$\bar{I}_y = 1.350 \times 10^6$ mm^4.
9.34 69.6 in^4, 11.64 in^4; 2.52 in., 1.029 in.
9.35 6123 in^4, 1361 in^4; 7.91 in., 3.73 in.
9.36 -27.3×10^6 mm^4.
9.37 $+20.45$ in^4.
9.38 -5.00 in^4.
9.39 $+1.561 \times 10^6$ mm^4.
9.40 10.17 in^4, 5.51 in^4; -5.96 in^4.
9.41 4.48×10^6 mm^4, 1.36×10^6 mm^4,
-1.15×10^6 mm^4.
9.42 $I_u = I_v = 5a^4/24$; $I_{uv} = -a^4/8$.
9.43 $\bar{I}_u = 0.267a^4$, $\bar{I}_v = 0.567a^4$, $\bar{I}_{uv} = -0.200a^4$.
9.44 $-25.7°$; 14.24 in^4, 1.437 in^4.
9.45 $-26.8°$; 4.86×10^6 mm^4,
0.982×10^6 mm^4.
9.46 $+23.2°$; 28.1 in^4, 5.14 in^4.
9.47 $-25.1°$; 83.2 in^4, 20.8 in^4.
9.48 10.17 in^4, 5.51 in^4; -5.96 in^4.
9.49 4.48×10^6 mm^4, 1.36×10^6 mm^4,
-1.15×10^6 mm^4.
9.52 $-25.7°$; 14.24 in^4, 1.437 in^4.
9.53 $-26.8°$; 4.86×10^6 mm^4, 0.982×10^6 mm^4.
9.56 (*a*) $I_{AA'} = mb^2/24$; $I_{BB'} = mh^2/18$.
(*b*) $I_{CC'} = m(3b^2 + 4h^2)/72$.
9.57 (*a*) $\frac{1}{6}mh^2$. (*b*) $\frac{1}{2}mh^2$.
9.58 (*a*) $\frac{1}{4}ma^2$, $\frac{1}{4}mb^2$. (*b*) $\frac{1}{4}m(a^2 + b^2)$.
9.59 $ma^2/20$.
9.60 $\frac{1}{12}\rho\pi a^2L\,(3a^2 + 4L^2)$, $\sqrt{(3a^2 + 4L^2)/12}$.
9.61 $m(2b^2 + h^2)/10$.
9.62 $m(a^2 + b^2)/5$.
9.63 1.305 kg $\cdot$ m^2; 197.1 mm.
9.64 1.208 lb $\cdot$ ft $\cdot$ s^2; 6.23 in.
9.65 (*a*) $4ml^2/3$. (*b*) $ml^2/12$. (*c*) $17ml^2/12$.
9.66 (*a*) $md^2/6$. (*b*) $2md^2/3$. (*c*) $2md^2/3$.
9.67 0.0767 kg $\cdot$ m^2; 84.9 mm.
9.68 0.0503 lb $\cdot$ ft $\cdot$ s^2; 3.73 in.
9.69 (*a*)1.575 lb $\cdot$ ft $\cdot$ s^2. (*b*) 2.714 lb $\cdot$ ft $\cdot$ s^2.
(*c*) 1.202 lb $\cdot$ ft $\cdot$ s^2.
9.70 (*a*) 7.82 kg $\cdot$ m^2. (*b*) 16.28 kg $\cdot$ m^2.
(*c*) 15.96 kg $\cdot$ m^2.
9.71 87.3×10^{-3} kg $\cdot$ m^2.
9.72 $I_{xy} = -0.00991$ lb $\cdot$ ft $\cdot$ s^2, $I_{yz} = I_{zx} = 0$.
9.73 $I_{xy} = -0.001199$ kg $\cdot$ m^2, $I_{yz} = I_{zx} = 0$.
9.74 $I_{xy} = -48.0 \times 10^{-6}$ lb $\cdot$ ft $\cdot$ s^2,
$I_{yz} = +48.0 \times 10^{-6}$ lb $\cdot$ ft $\cdot$ s^2, $I_{zx} = 0$.
9.75 $I_{xy} = I_{zx} = 0$, $I_{yz} = -0.0932$ lb $\cdot$ ft $\cdot$ s^2.
9.76 $I_{xy} = 5.43$ kg $\cdot$ m^2, $I_{yz} = I_{zx} = 0$.
9.77 $I_{xy} = 0.502 \times 10^{-3}$ kg $\cdot$ m^2,
$I_{yz} = -1.005 \times 10^{-3}$ kg $\cdot$ m^2,
$I_{zx} = -1.256 \times 10^{-3}$ kg $\cdot$ m^2.
9.78 9.73 kg $\cdot$ m^2.
9.79 1.892 lb $\cdot$ ft $\cdot$ s^2.
9.80 10.88×10^{-3} lb $\cdot$ ft $\cdot$ s^2.

9.81 (*a*) $\frac{1}{12}m(a^2 + b^2)$. (*b*) $\frac{1}{12}m(b^2 + c^2)$.
9.82 (*b*) $ma^2/20$, $17ma^2/40$.
9.83 $7ma^2/40$.

CHAPTER 10

10.1 225 lb←.
10.2 450 lb · in.↓.
10.3 75 N↓.
10.4 $M = 2Wl \sin \theta$.
10.5 $\frac{1}{2}Pb \cos \theta$.
10.6 $(Pl/a) \cos \theta \sin^2 \theta$.
10.7 $(P \cos \theta)/\sin (\theta/2)$.
10.8 $\frac{1}{2}Pl \tan \theta$.
10.9 $\frac{1}{2}Pl \tan \theta/(1 - \mu \tan \theta)$.
10.10 32.3°.
10.11 $M = 2Pl(\tan \theta - \mu)$. For $\mu \geq \tan \theta$, mechanism is self-locking.
10.12 13.50 in.
10.13 25.98 N · m.
10.14 541 N.
10.15 33.6°.
10.16 24.8°.
10.17 23.9 kN↗.
10.18 (*a*) 12.5(*HE*). (*b*) 40.5°.
10.19 6.93*W* *comp*.
10.20 3.49*W*.
10.21 $(Pl/r) \cos^2 \theta$.
10.22 $(M \cos^2 \theta)/(r \sin \theta)$.
10.23 50 lb.
10.24 30 lb.
10.27 $\theta = 45°$, unstable; $\theta = -135°$, stable.
10.28 78.7°, unstable; −101.3°, stable.
10.29 0°, stable; 60°, unstable.
10.30 $\sqrt{2} \sin \theta - \tan \theta = W/ak$.
10.31 10.9°; 39.5°.
10.32 $P < k(l - a)^2/2l$.
10.33 $\tan \theta(1 - \cos \theta) = W/4lk$.
10.34 51.3°.
10.35 (*a*) $\frac{1}{2}Pb \cos \theta$. (*b*) 36.9°, unstable.
10.36 (*a*) 200 lb. (*b*) 10 in.